Understand Amplifiers

Understand Amplifiers

Owen Bishop

Newnes

OXFORD BOSTON JOHANNESBURG MELBOURNE NEW DELHI SINGAPORE

Contents

1

Amplifying signals

The purpose of an electronic amplifier is to make an electronic signal bigger without affecting it in any other way. Before we can consider exactly what is implied by this statement, we need to understand the nature of a typical electronic signal. The first point is that a signal is a *current*, which varies in size and possibly in direction from instant to instant. When the signal represents a sound, the current changes too rapidly to be measurable with an ammeter. One way of examining it is to pass it through a resistor and connect an oscilloscope across it (Fig. 1.1). According to Ohm's Law, a current flowing through a resistor causes a potential difference to appear between its two ends and that potential difference is exactly proportional to the current. The circuit in Fig. 1.1 converts a varying current into a varying voltage difference. The oscilloscope responds rapidly to the changing voltage and produces its display. If the sound is that of a clarinet, for example, we might see a line with a very complicated shape, such as the one in the figure.

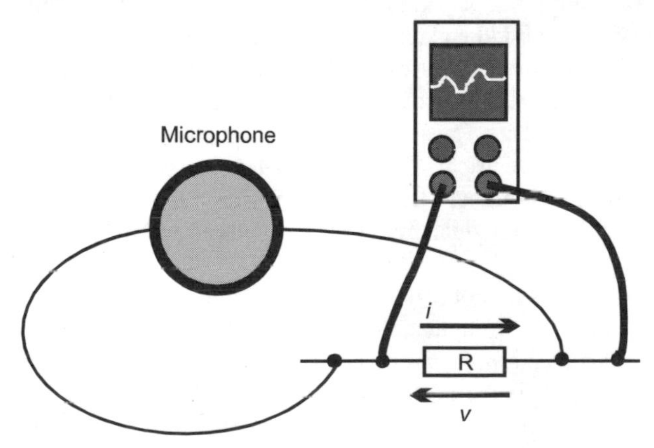

Figure 1.1 *As current varies, there is a varying voltage across the resistor, which is displayed by the oscilloscope.*

It is easier to understand signals if we look at one of the simplest signals, the *sine wave* or *sinusoid* (Fig. 1.2). This is the kind of signal that we can get from an electronic signal generator. Note that the signal is repeated over and over again. It is a *periodic* signal, and the length of the *period* of the signal in Fig. 1.2 is 1 millisecond (1 ms). The period *P* is one way of expressing how often the signal repeats.

Another way is to state how many times it repeats in 1 second. We call this its *frequency*, which is measured in *hertz*. The signal of Fig. 1.2 repeats

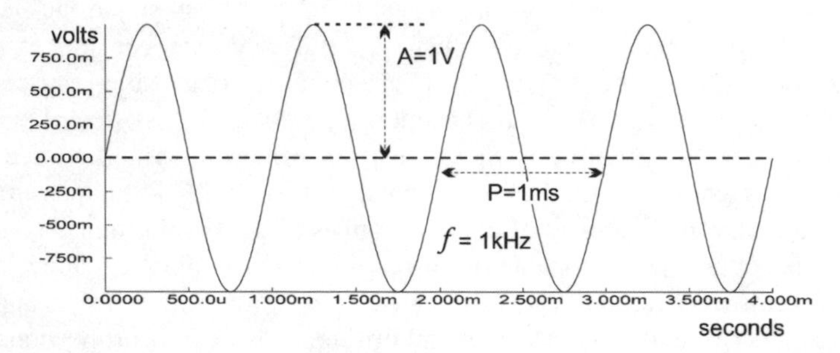

Figure 1.2 *Sinusoids are the basic components of all periodic waveforms.*

itself 1000 times in 1 second so its frequency is 1000 hertz, shortened to 1000 Hz, or 1 kilohertz, or 1 kHz. A 1 kHz note has about the same pitch as the second C above middle C on a piano. In terms of sound, a note with low frequency has low pitch. The lowest note that most people can hear has a frequency of about 30 Hz. At the other extreme, the highest note we can hear is about 20 kHz. These two values are the limits of the *audio-frequency range*. Electrical signals in radio transmitters and receivers generally have much higher frequencies, measured in megahertz (1 MHz = 1 million hertz) or even gigahertz (1 GHz = 1 thousand megahertz).

Musical instruments do not produce such a simple signal as Fig. 1.2 but, as we shall see later, the complicated signals of musical instruments and other sources of *periodic signals* can be considered to be made up of many sine waves combined. It is the particular assortment of sine waves of different frequencies that gives each kind of musical instrument its characteristic sound, or *timbre*.

In the discussions in this book we usually consider how a circuit responds to a signal that is a pure sine wave of one particular frequency. But all periodic sounds are made up of a mixture of sine waves of different frequencies, and the same conclusions apply to each and every frequency present.

Sine waves

When plotted as a graph with time along the x-axis and current along the y-axis, a sine wave signal has exactly the same shape as the curve we obtain when we plot the sine of an angle against the angle itself. Although we may think of sines as belonging to a rather abstruse branch of mathematics, sines play an important part in our lives. The motion of a child on a swing, the vibrations of the sound of a flute and the oscillations of an electronic circuit can all be expressed in terms of sines.

The *amplitude* of the signal, is defined in Fig. 1.2. In the figure, A is constant but with other signals it may increase or decrease with time. In terms of sound, this produces a tone that gradually becomes louder or softer.

Similarly, the *frequency* may change with time, as in Fig. 1.3. In terms of sound, increasing frequency produces a rising pitch, as when a whistling kettle starts to boil. Decreasing frequency produces a falling pitch, as when

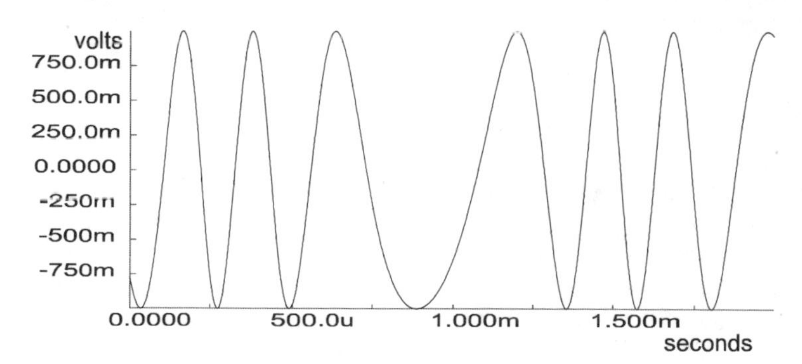

Figure 1.3 *The frequency of this signal varies in time. We say that it is frequency modulated. In this example, the frequency is modulated by a relatively large percentage to make the drawing clearer, but usually the percentage modulation is much less.*

an electric train slows down and stops at a station.

Periodic signals

Fig. 1.4 shows four sinusoidal signals. The one with lowest frequency and amplitude is the *fundamental*. This has frequency f (here, $f = 1$ kHz). The other signals have frequencies that are multiples of the fundamental (in this case, $2f$, $4f$ and $6f$, but with smaller amplitudes). They are called the *harmonics*. Combined, these signals produce a signal of complicated shape (Fig. 1.5). This signal is periodic, with a frequency equal to that of the fundamental but has a distinctive timbre, perhaps characteristic of a particular musical instrument.

Even signals of very angular shape, such as square waves and sawtooth waves, are found to consist of a combination of sine waves, the fundamental and various harmonics.

Current amplification

A varying current i_{IN} is fed to the input of an amplifier. A varying current i_{OUT} comes from the output. At every instant, i_{OUT} is proportionate to i_{IN}. We define the ratio between input and output currents at any instant as the *current gain*, A_i of the amplifier, where:

$$A_i = \frac{i_{OUT}}{i_{IN}}$$

For amplification, as usually defined, the value of A_i is greater than unity. There are instances when we require unity gain or even gains of less than unity. We often have negative gain, in which the output signal is the inverse of the input signal. But these are special cases of the principle of amplification and do not affect the general discussion, in which we assume that amplification 'makes a signal bigger'.

A *current amplifier* has many applications. For example, an amplifier driving a motor in a robot may need to supply several amps of current. Yet only a few milliamps are required to drive the controlling circuits. Large currents usually mean large power ratings. A public address system, for

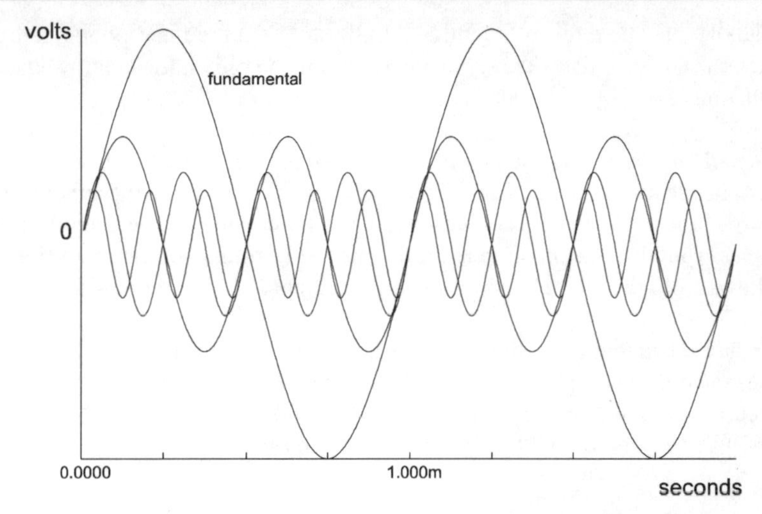

Figure 1.4 *These four signals comprise the fundamental sinusoid (1 kHz) and three harmonics (2, 4 and 6 kHz) with progressively smaller amplitude.*

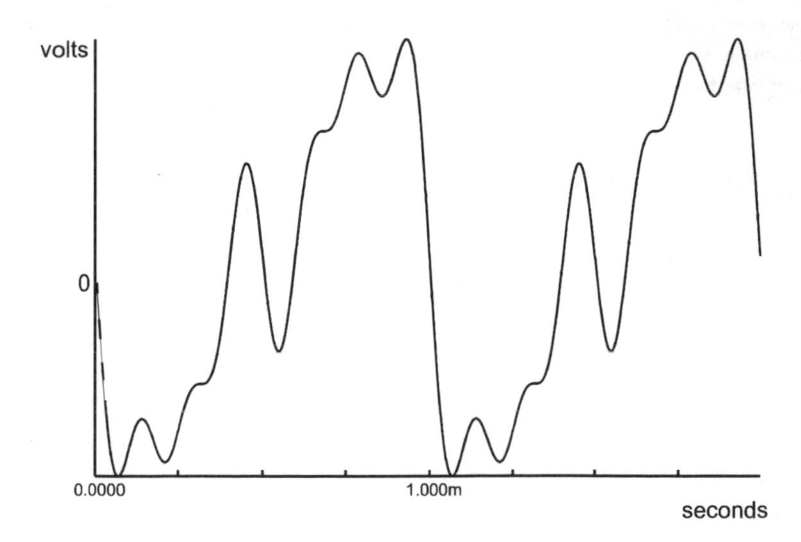

Figure 1.5 *The sum of the four signals of Fig. 1.4 produces the complicated signal shown above, which has frequency 1 kHz. Conversely, this signal can be analysed into the four sinusoids of Fig. 1.4.*

example, may be rated at several hundred watts. Yet the microphone or disc player which originates the signals may operate on only a few microwatts or milliwatts.

A small current enters the amplifier and a larger one leaves it. Current is a flow of electric charge. The larger the current, the larger the amount of charge flowing. This extra charge can not be created in the amplifier. It comes from the current supplied to the power terminals of the amplifier. It follows from this that all amplifiers need a supply of electric power.

We have seen that a resistor can be used to convert a current into a voltage difference. If an amplifier has a resistor in its input circuit, or in its output circuit or in both, we can measure the input and output voltages produced. Then we can find the *voltage gain* A_v of the amplifier:

$$A_v = \frac{v_{OUT}}{v_{IN}}$$

Voltage gain is important when we have a signal source such as a microphone which produces a signal of only a few millivolts amplitude, or when we have a radio signal of only a few microvolts amplitude picked up by an antenna. For these signals to be of any use to us, their voltage must be amplified.

Ideal amplifier

We are now able to specify the characteristics of an ideal amplifier. From our definition at the beginning of this chapter, its most important characteristic is its *gain,* by which we usually mean *voltage gain*. For an ideal amplifier:

- A_v or A_i are constant whatever the value of v_{IN} or i_{IN}. In other words, the gain of the amplifier must not be affected by the amplitude of the signal.

- Gain must not be affected by frequency. Signals of all frequencies are to be amplified by exactly the same amount.

- The amplifier must not add noise to the signal and, if possible, should reduce or remove any noise that is already in the signal.

- The ideal amplifier does not pick up noise from outside sources.
- The performance of an amplifier should be unaffected by ambient temperature. Temperature stability is very important in amplifiers that have to operate in extreme environments, such as those used in aircraft and satellites.

- The performance of an amplifier must be stable over a period of time. The values of certain components (electrolytic capacitors are a prime example) change with age. The circuit design must allow for this and compensate for it as far as possible.

As might be expected, no *real* amplifiers are ideal , but there are techniques for minimising their imperfections. We shall discuss these as we look at real amplifiers in the remaining chapters of this book.

Noise

Noise is any unwanted signal that has become superimposed on the wanted signal. One category of noise is the result of random motion of the charges in the circuit of the amplifier itself. This shows up in the output of an audio system as various types of hissing sound. In a control system it may result in erratic responses or uneven motion. These kinds of noise and how to avoid them are discussed in more detail later. Another kind of noise is an electromagnetic signal entering the circuit from its surroundings, either directly or through the mains supply. Once these signals have entered the amplifier circuit they are added to the signal which is being amplified and appear, usually in amplified form, at the output. This type of noise may be caused by the switching on or off of heavy loads in nearby equipment, by lightning strikes and similar electrical events. Noise may even consist of radio signals picked up from nearby powerful transmitters. This type is more often termed *interference*.

Test yourself

(Answers to numerical questions are in the Appendix)

1. What is the main function of an electronic amplifier?

2. What is meant by an 'electronic signal'?

3. What is the basic type of signal of which all periodic signals are composed?

4. Define the frequency of a signal. In what unit is this expressed?

5. A complicated periodic signal has a frequency of 750 Hz. What is the frequency of (a) the fundamental, and (b) the first harmonic?

6. Define the current gain of an amplifier.

7. Name a type of component that is well-known for its instability.

8. What term do we use to describe an unwanted signal that is superimposed on a wanted signal?

9. Name some sources of such unwanted signals, stating which originate inside the amplifier and which originate outside.

10. List the factors by which an ideal amplifier should *not* be affected.

2
A simple amplifier

Amplifiers are the basis of many types of electronic circuit. When we first think of amplifiers we think of the many types of audio amplifier that abound in the home, from the amplifier in the stereo or CD player to those in the video recorder, not forgetting the more humble amplifiers we find in a telephone, a fax machine or a microwave oven (amplifying the 'beep'). Then we consider amplification of other types of signal. Cameras, security systems, and dishwashers provide several instances of amplification. But signal amplification is not the only use for an amplifier, for most oscillators include some kind of amplifier. The 'beeps' and many other sounds and flashing lights which produce the output of so much domestic and industrial equipment, nearly always depend on an amplifier. So do the oscillators that form part of radio and TV circuits. Coming back to the stereo player again, we find amplifiers as an essential part of the audio filters that assure the high quality of the sound. But no matter how many and varied the applications, the fundamentals of amplification are relatively few and easily understood. This chapter deals with one particular type of amplifier which illustrates many of the essential features.

We begin by looking at an amplifier that is based on a single transistor. There are several different kinds of transistor that we could use. The easiest to understand is the *field effect transistor*, or FET. From the several different kinds of FET that are manufactured, we have chosen the MOS-FET, which is short for Metal-Oxide-Silicon FET. Before we can see how a MOSFET amplifier works, we need to look at the MOSFET itself.

Fig. 2.1 shows a section through a typical MOSFET. The transistor is constructed on a block of p-type silicon, known as the *substrate*. At the upper surface there are two regions, or *wells*, of n-type silicon. A thin film of metal, usually aluminium, is deposited over these wells during manufacture to make connections to the terminals of the transistor, the *source* and the *drain*. Between the wells, the surface of the silicon substrate is oxidised. Silicon oxide is a very effective insulator and on this is deposited another film of metal, called the *gate*. Because of the insulation, it is not possible for current to flow between the gate and the rest of the transistor.

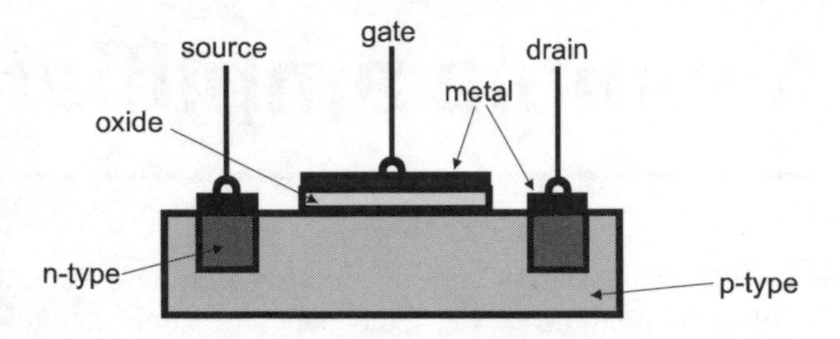

Figure 2.1 *A MOSFET consists of metal, oxide and semi-conductor. In an n-channel MOSFET, the semiconductor is p-type, with two wells of n-type*

Fig. 2.2 shows how the transistor works. The substrate and source are wired together during manufacture (the substrate can have a separate terminal, but this is unusual). In use, both are normally connected to the 0 V or most negative supply. The drain is connected to the positive supply. No current can flow through the transistor at this stage because current can not flow when an n-type region is positive of a p-type region (more about this in Chapter 5).

The situation changes if we make the gate a few volts positive of the source. The positive charge on the gate electrode creates an *electric field*. The effect of this field is to repel positive charges (holes) from the region of the

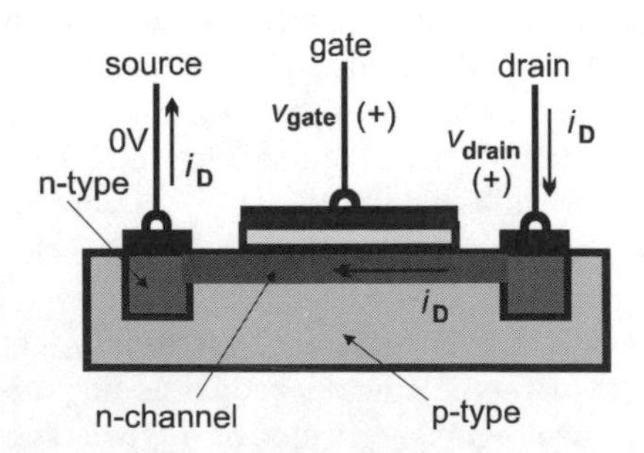

Figure 2.2 *Now that the gate has a positive charge, the effect of its field is to repel positive charges (holes), creating an electron-rich channel through which (conventional) current can flow from drain to source.*

N-type and P-type semiconductors

The ability of a semiconductor to conduct electric charge is increased if it is *doped.* This means replacing some of the atoms of silicon, germanium or other semiconducting material in the crystal with atoms of a different kind. For example, we may dope a block of silicon by diffusing a small proportion of antimony into it. Adding the antimony increases the conductivity by providing an *increased* supply of free electrons. Electrons carry *negative* charge, so a semiconductor that has been doped to provide extra electrons is known as *n-type* semiconductor.

Another way of doping is to add a dopant that *lacks* electrons. For example, silicon may be doped with boron. This creates vacancies (or *holes*) in the crystal structure. Under the influence of an electric field, electrons flow from negative to positive, jumping from hole to hole (Fig. 2.3). As electrons flow in one direction, this leaves empty holes upstream. The effect of this is that the holes appear to 'travel' in the opposite direction, from positive to negative. The effect is the same as if we had *positive* charges flowing in the semiconductor. We call this *p-type* semiconductor.

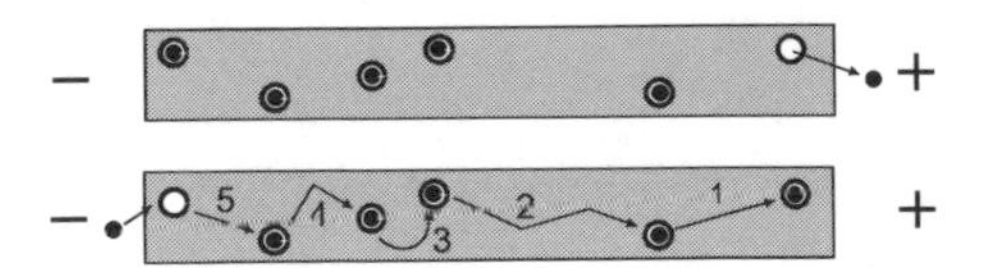

Figure 2.3 *Conduction by holes. Top: an electron escapes from an atom, leaving a vacancy, or 'hole'. Bottom: 1-5 show the sequence of electrons jumping from hole to hole. Electrons move from negative to positive, but holes apparently move from positive to negative. They act as positive charge carriers.*

substrate close to the gate. Or you can think of it as attracting electrons into the region and so filling up the holes. The result is an increase in the number of electrons in that region, converting the p-type semiconductor to n-type. When this has happened, a continuous region of n-type exists from the source to the drain. Current, known as the *drain current*, flows through the transistor. The bigger the charge on the gate, the more electrons become available and the more drain current flows.

Current and charge

An electric current is a transfer of electric charge from a point where potential is relatively high (positive) to a point where potential is relatively low (negative). This is the conventional view of current, which we use throughout this book. In fact the transfer of charge is more complicated than this because of the two kinds of charge. Positive charge carriers (positive ions, holes) flow from high to low potential (positive to negative). Negative charge carriers (negative ions, electrons) flow from low to high potential (negative to positive).

In descriptions of the working of transistors remember that the flow of electrons is in the opposite direction to 'conventional' current.

Input and output

The extent to which the drain current depends on the drain-source and gate-source voltages is measured by using a circuit such as that in Fig. 2.4. One of the measurements we can make shows how drain current varies with the voltage of the gate. This graph is known as the *forward transfer characteristic* (Fig. 2.5). It is called this because it shows how changes in input are transferred forward across the transistor and bring about changes in output. We use the symbol i_D for the drain current, and the symbol v_{GS} for the gate-source voltage.

Below a certain value of v_{GS}, known as the *threshold voltage* V_T, there is no current through the transistor because the field effect is insufficient to create an n-type channel. But, as v_{GS} is gradually increased above the threshold, the channel becomes wider and the amount of current increases. There is an upper limit to this, when behaviour changes, but we are concerned only with the region shown in the figure.

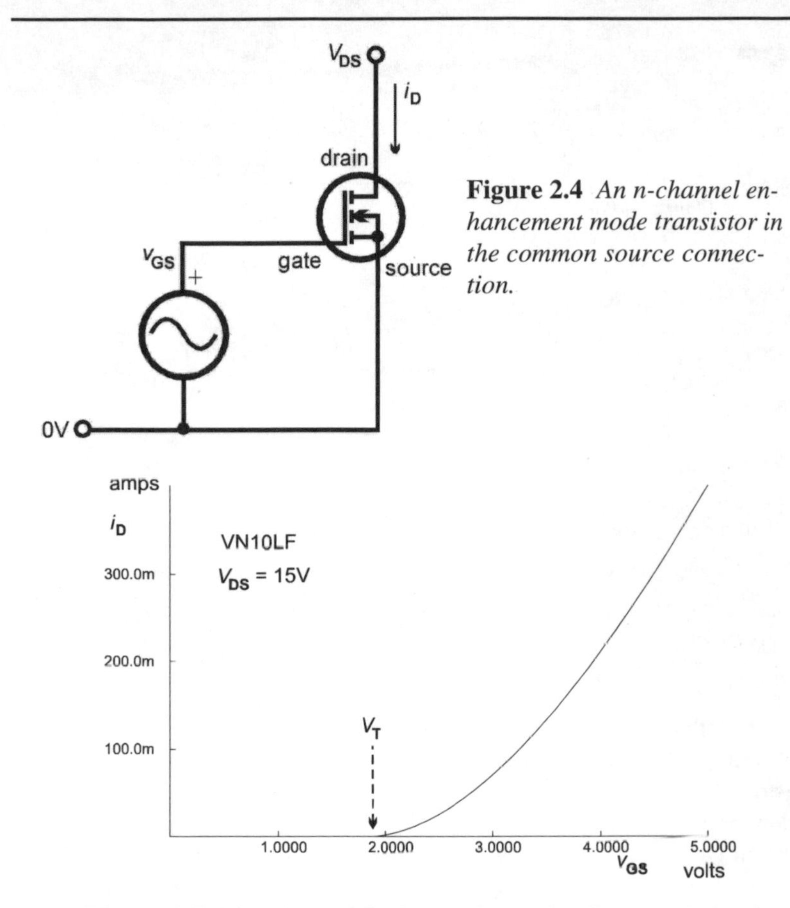

Figure 2.4 *An n-channel enhancement mode transistor in the common source connection.*

Figure 2.5 *The slope of the forward transfer characteristic of an n-channel MOSFET increases with increasing v_{GS}.*

Plotting MOSFET responses

Another way of investigating the behaviour of a MOSFET is to plot a series of curves as in Fig. 2.6. This is known as the *output characteristic*. Looking at any one curve of this family, we can see how the drain current (the output) increases with increasing drain-source voltage, V_{DS}. There is a steep increase at first, when current is proportional to voltage. In this region, known as the *resistive* or *ohmic* region of the curve the FET behaves like a

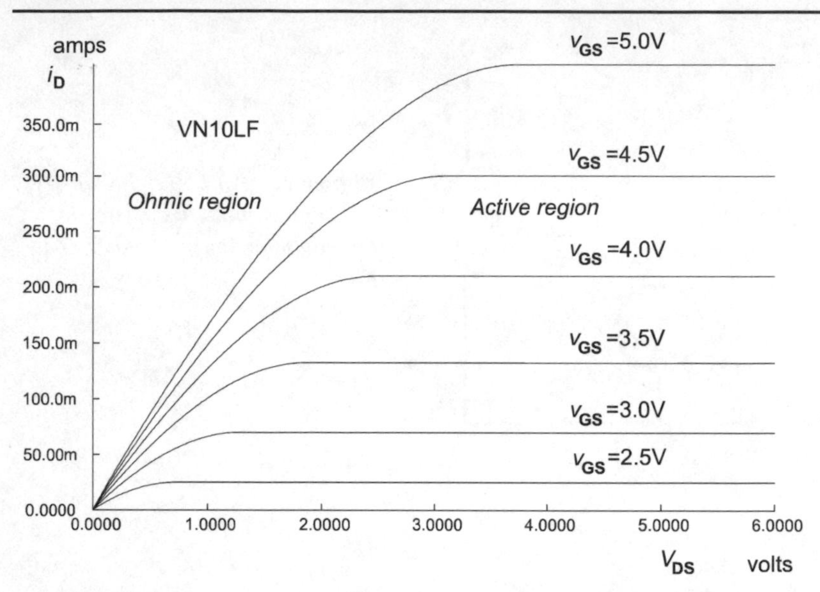

Figure 2.6 *The output characteristic of an n-channel enhancement MOSFET levels off in the active region at values proportional to gate-source voltage.*

resistor, with its resistance being controllable by the value of v_{GS}. In certain applications, the FET is operated in this region as a voltage-controlled resistor. Beyond the resistive region, the curve gradually flattens out. This region of the characteristic is known as the *active region*. Comparing different curves, the level eventually reached in the active region depends on the gate-source voltage v_{GS}. This is to be expected from the results in Fig. 2.5. In an amplifier, we aim to operate the transistor in the active region, where output (i_D) increases with increasing input (v_{GS}), and is relatively unaffected by V_{DS}. But note that the horizontal sections of the curves are more widely spaced with increasing v_{GS}. This is because i_D increases more and more rapidly with increase of v_{GS}, as we saw in the upward sweeping curve in Fig. 2.5.

The *input* to the transistor is a *voltage* applied between gate and source. The output is a *current* of electrons flowing from source to drain. The source is involved in both the input and output sides. It is *common* to both sides and, for this reason, we call this *common-source connection*. Other connections are possible, as we shall see in the next chapter.

When designing a circuit, a question we may ask is ‘By how much does the

MOSFET types

The MOSFET just described in the main text conducts through an n-type region (an *n-channel*) starting at the threshold with no channel and gradually increasing the channel width with increasing gate voltage. In other words, the transistor operates by enhancing the channel. For these reasons the transistor is called an n-channel *enhancement* MOSFET. The n-channel enhancement MOSFET is the most commonly used type.

Other sorts of MOSFET are made, of which the next most commonly used is the p-channel enhancement type. The substrate of this is n-type silicon, and the drain and source regions are p-type. A *negative* voltage applied to the gate creates a p-type region between source and drain.

The other kind of MOSFET is the *depletion* type. This type is manufactured with a continuous conductive region (either n-type or p-type) between source and drain (Fig. 2.7). The channel is fully conducting with 0 V gate input. Given a negative voltage at the gate, electrons are repelled, so depleting the channel of charge carriers, and reducing i_D.

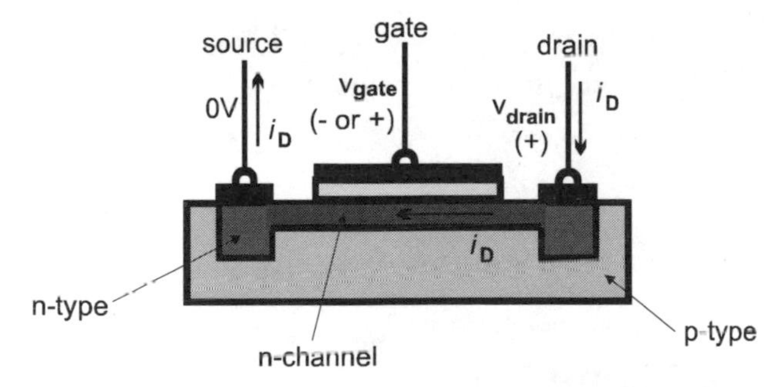

Figure 2.7 *When the gate of a depletion n-channel MOSFET is made negative of the source, holes are attracted, reducing the width of the channel and so reducing conduction. When the gate is made positive, the channel is made wider, which increases conduction.*

current increase when the gate-source voltage is increased by a given small amount?' (We assume that the drain source voltage is being held constant). Fig. 2.5 shows that there is no one answer to this. The increase in current for a given increase in voltage is the slope or gradient of the curve, but this depends on what part of the curve we use to get the figures. The slope increases with increasing v_{GS}. At any particular point on the curve we can measure the slope by increasing v_{GS} by a very small amount and measuring the small increase in i_D. The slope of the curve at that point is:

$$\frac{\text{increase in } i_D}{\text{increase in } v_{GS}}$$

This ratio is called the *transconductance*. The name is short for 'transfer conductance'. Transconductance is a kind of conductance because it consists of a current divided by a voltage, but it is 'transferred' across the transistor because it relates the *input* voltage to the *output* current. Its symbol is g_m. The unit of transconductance is the siemens (symbol, S), equivalent to amps per volt. Transconductance varies from one type of MOSFET to another, ranging between 100 mS and 20 S, but varies with i_D in a given transistor, as explained above.

Keeping up?

(all questions assume an n-channel enhancement MOSFET)

1. What does MOSFET stand for? Which parts are made of metal, which of oxide and which of silicon, where is the field, and what is its effect?
2. By what terminal do electrons enter the MOSFET? By what terminal do they leave?
3. If a MOSFET is working in the active region, has a g_m of 0.8 S and the gate voltage is increased by 0.15 V, what is the change in its drain current?

MOSFET amplifier

Having looked at the way a MOSFET operates in the common-source connection, we are ready to build an amplifier around it. In many ways, this amplifier is the basic transistor amplifier. We will begin by describing its simplest form (Fig. 2.8) and then improve its performance by a number of modifications.

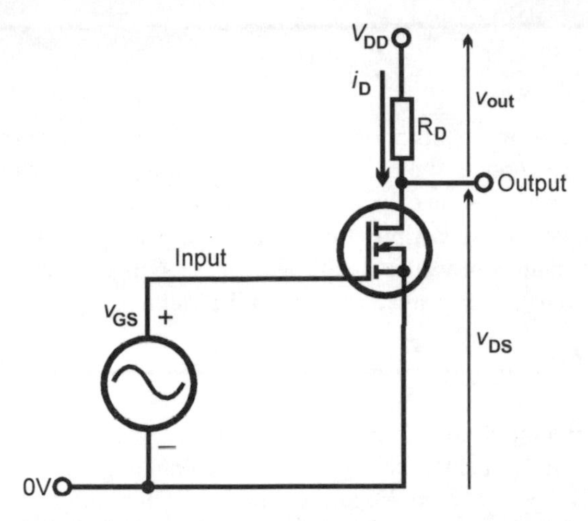

Figure 2.8 *A drain resistor converts the output current into an output voltage. Note: the supply is now labelled V_{DD}.*

Using Ohm's Law

A resistor is useful for *converting current to voltage*. When a current I is passed along a resistance R, a voltage (potential) difference V is generated between its two ends. According to Ohm's Law:

$$V = IR$$

The law holds true for every instant of time, so that changes of current are instantly converted to changes of voltage.

Second only to the transistor itself, the drain resistor R_D is the next most important component of the amplifier. The resistor acts as a current-to-voltage converter. As input voltage (v_{GS}) varies, drain current (i_D) varies and this in turn makes the voltage across the drain resistor vary. We feed a varying voltage into the amplifier and obtain a varying voltage as its output. The next step is to find the voltage gain, A_v, as defined in Chapter 1.

Symbols

In the descriptions in this book we express fixed, average or quiescent values by all-capitals symbols such as V_{DS}, I_D, R_S. The instantaneous values of quantities that vary are represented by lower-case letters with capital-letter subscripts, for example, v_{DS}, v_{OUT}, i_D. We express small fluctuations in values by all-lower-case symbols such as v_{in},v_{ds}, i_d. For example, v_{ds} is a small *change* in the value of v_{DS}.

In Fig. 2.8, the input voltage goes directly to the gate, so v_{IN} is the same thing as v_{GS}. Given a small change of input voltage v_{in}, the corresponding change of drain current is i_d. We know that, by definition, $i_d/v_{gs} = i_d/v_{in} = g_m$. Rearranging the equation gives: $i_d = g_m \times v_{in}$. But, if the drain resistor is R ohms, the change of voltage across it is $v_{out} = i_d \times R$.

Combining these two equations gives: $v_{out} = g_m \times v_{in} \times R$.

Output is proportional to both transconductance and the size of the resistor.

The output equation can be rearranged to give an equation for the voltage gain, A_v as defined in Chapter 1:

$$A_V = \frac{v_{out}}{v_{in}} = g_m \times R$$

Voltage gain is directly proportional to the transconductance and to the value of the drain resistor. Transconductance varies with the type of MOSFET. In Fig. 2.6, for $v_{GS} = 4$ V, the transconductance is approximately equal to 150 mV per volt, or 150 mS, or 0.15 S. Describing it in another way, we say that when v_{GS} equals 4, the *gradient* of the curve is 150 mV per volt. If R is 100 Ω, then $A_v = 0.15 \times 100 = 15$. We could obtain a larger gain by using a larger resistor or a MOSFET with higher transconductance.

Increasing input voltage results in an increasing voltage *across* the drain resistor. Since this resistor is connected to the supply voltage at one end and the output voltage is taken from its *other* end, an increasing voltage across the resistor means a decreasing voltage at the output. The amplifier is an *inverting amplifier.*

The inverting action described above is a property of the *transistor* when connected as in Fig. 2.8, that is, in the common-source connection. It does not necessarily apply to a common-source *amplifier* consisting of a transistor plus various resistors and capacitors. The action of the capacitors may introduce delay into the amplification process so that the output is no longer the instant-by-instant inverse of the input.

Gate biasing

The next step in building an amplifier is to provide bias for the gate so that a steady current flows through the transistor, even when there is no signal present. This holds the gate voltage comfortably above the threshold voltage, and in the active region (Fig. 2.6), allowing the input signal to produce a fall at the gate as well as rise. The standard method for providing bias is to use two resistors as a *potential divider* (see box). A circuit in which this principle is applied to biasing a transistor is pictured in Fig. 2.9. We use the equation in the box to calculate the values required. Any pair of values in the correct proportions would produce the biasing voltage we want. The MOSFET takes almost no gate current, so the situation illustrated in Fig. 2.11 does not arise. Resistances can be high, in the order of tens of kilohms. If, for example, the supply voltage is 9 V and we use the standard resistor value of 560 kΩ for R_{G1} and 390 kΩ for R_{G2}, the bias voltage we obtain is 9 × 390/950 = 3.7 V. This puts the action of the transistor close to the 3.5 V curve of Fig. 2.6.

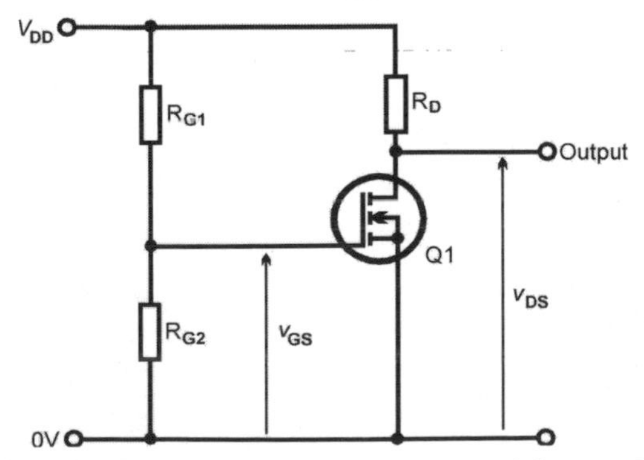

Figure 2.9 *The first stage in making the amplifier ready to receive an input signal.*

Potential divider

This consists of two resistors in series (Fig. 2.10). The same current flows through both resistors. If the total resistance is R_T, then $i = V/R_T = V/(R_1 + R_2)$, $i = v_1/R1$ *and* $i = v_2/R_2$

Looking at R_2, we can say that $v_2 = i \times R_2 = V/(R_1 + R_2) \times R_2$. Rearranging this equation into its more usual form:

$$v_2 = V \times \frac{R_2}{R_1 + R_2}$$

A similar equation is found for v_1. In words, the voltage across *one* of the resistors equals the total voltage, multiplied by the value of that resistor, and divided by the total resistance.

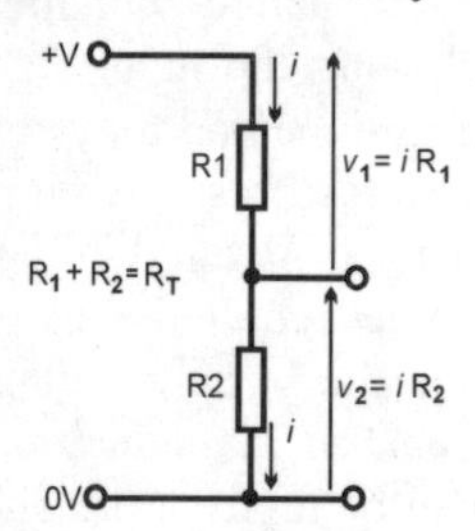

Figure 2.10 *A potential divider network divides the input potential in proportion to the resistance.*

The calculation assumes that all the current flowing through R1 flows on through R2. If a portion of the current flows out at their junction (Fig. 2.11) this in effect places the load in parallel with R2. This reduces the voltage across R2, lowering the voltage at the junction. The more current flowing out, the bigger the drop in voltage. In the biasing of a MOSFET, the current flowing out of the divider is so tiny that the divider resistors can be of high value.

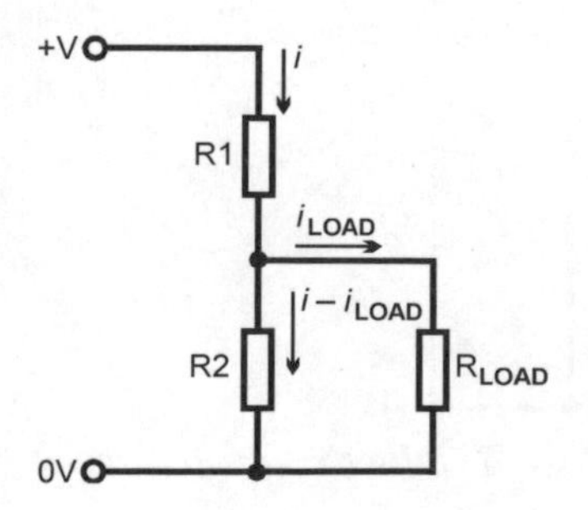

Figure 2.11 *When a load is connected to the potential divider, less current flows through R2 and the output voltage falls.*

Coupling

The next task is to arrange for the input signal to be fed into the amplifier and the output signal to be taken from it. If the signal source has an average no-signal level of a few volts (say, 4 V) it can be wired directly to the gate and we may not need the biasing resistors discussed in the previous section. But most often this will not be the case. More often the average no-signal voltage (or *quiescent voltage*) will be 0 V. As an example, a crystal microphone has an output of 0 V when it is receiving no sound. When it is receiving sound, its output signal ranges between a few millivolts positive and a few millivolts negative.

To allow the signal source and the gate of the transistor to rest at different quiescent voltages, we *couple* them. This can be done without making a direct electrical connection if we use either an inductor or a capacitor.

When we use an inductor for coupling, we wind two coils of fine wire on the same former. Often there is a core of ferromagnetic material such as ferrite in the former to increase the magnetic linkage between the two coils. The inductor functions in the same way as a transformer. One coil is wired into the output circuit of one stage (which might be a sensor, or the first stage of a multi-stage amplifier). The other coil is wired into the input circuit of the next stage (see Fig. 7.12). The signal passes from one coil to the other. Inductors tend to be bulky, heavy and relatively expensive, so their use is generally avoided in low-cost and portable equipment. Their main use is in high-frequency amplifiers, where their inductance may be of advantage in tuned circuits.

When we use a capacitor for coupling, the linked output and input circuits are connected one to each plate. There is no electrical connection between them, for the plates of the capacitor are separated by the dielectric, which is an insulator. One side may be at 0 V and the other at, say, +4 V. Signals fluctuating a few millivolts positive and negative of 0 V are transmitted through the capacitor, making the gate voltage sweep a few millivolts above and below +4 V.

We bring another factor into the circuit when we introduce capacitor coupling. The capacitor and the biasing resistors act as a highpass filter. If the biasing resistors are 560 kΩ and 390 kΩ, as suggested earlier, their parallel equivalent between the input line and ground line is 230 kΩ. If a 100 nF capacitor is used for coupling, the cut-off point of the highpass filter so formed is:

$$f_c = 1/2\pi RC = 6.9 \text{ Hz}$$

Signals of frequency 6.9 Hz or below are reduced to half power or less. This cut-off frequency is below the audio range, so we do not lose any signals of interest. A coupling capacitor is also needed at the output of the amplifier to couple it to the next stage. Here the highpass filter is formed by the capacitor and the drain resistor. If this has a resistance of 10 kΩ and the capacitor is 1 μF, the cut-off point is 16 Hz, which is suitable for audio applications.

It may be that the amplifier is *intended* to act as a filter, so that it does not, for example, amplify any 50 Hz mains interference that it accidentally picks up. Or maybe it is a radio-frequency amplifier intended to operate within the narrow frequency range of its carrier and sideband frequencies. In any such cases we calculate capacitor values accordingly, taking resistances into account. On the other hand, if the amplifier is intended to pass a wide range of frequencies, we take care to choose capacitor values that allow this to happen.

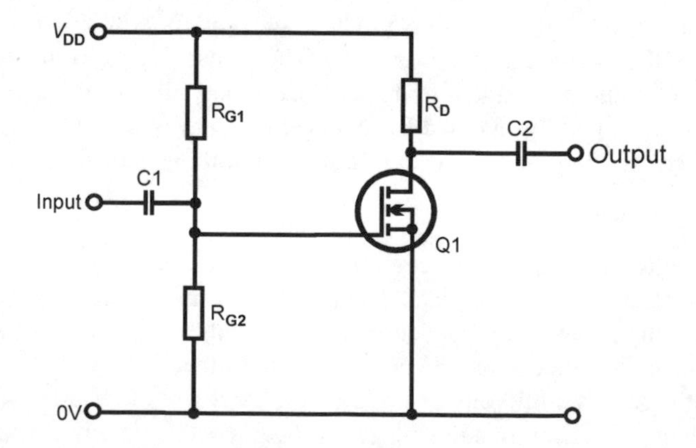

Figure 2.12 *Capacitor coupling allows signals to pass into and out of the amplifier, even though the quiescent voltages may not equal those in the circuits to which it is connected.*

Filters

In the electronic sense, a filter is a circuit which passes signals of one range of frequencies, more readily than those of other frequencies. A filter circuit may be quite complicated but even a single resistor and capacitor may act as a filter. The two most common filter circuits are contrasted in Fig. 2.13. To understand what they do, consider what happens when a signal is applied that consists of a mixture of different frequencies. In the lowpass filter, the capacitor passes high frequencies better than low frequencies, so the high frequencies are shunted to ground and mainly low frequencies appear at the output.

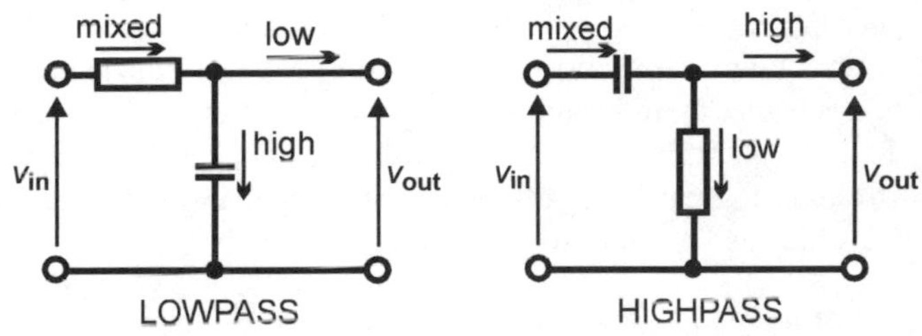

Figure 2.13 *The simplest filter circuits are built from a capacitor and a resistor.*

In the highpass filter, the capacitor again passes the high frequencies best. With the low frequencies that do pass the capacitor, there is time for the capacitor to charge or discharge through the resistor. The low frequencies are therefore damped. To put it another way, we can say that the low frequencies are conducted to ground and only the high frequencies appear at the output.

A filtering action also occurs with inductor coupling, the inductor and resistors forming a lowpass filter.

Keeping up?

(All questions refer to a common-source amplifier)

4. What is the function of the drain resistor?
5. If an amplifier has a 470 Ω drain resistor, and the g_m of the transistor is

25 mS, what is the voltage gain?

6. An amplifier operating on a 15 V supply has bias resistors 56 kΩ and 18 kΩ. What is the quiescent gate voltage?

7. Why do we usually need input and output capacitors in an amplifier? What effect may these have on signals entering or leaving the amplifier?

Input resistance

We now have to consider what happens when a signal current enters this circuit or leaves it. The signal might be the small current generated by a crystal microphone or picked up from a probe attached to the skin of a patient, so we do not wish to lose a significant portion of the signal during its transfer. Whether there is an input capacitor to couple the signal source to the gate of the amplifier makes little difference because the capacitor acts as a short-circuit to signals of sufficiently high frequency. In other words, the highpass filter lets them through.

As shown in Fig. 2.14, some of the signal entering an amplifier is shunted to ground instead of being amplified. Exactly which way it goes depends on the kind of network used for biasing. With a network of two resistors, as in Fig. 2.12, part of it flows along R_{G2} to ground. It is also partly shunted through R_{G1} to the positive supply rail but, since the positive and ground rails are connected together through the low resistance of the power supply, the signal is lost whichever way it goes. The result is that, in the input stage

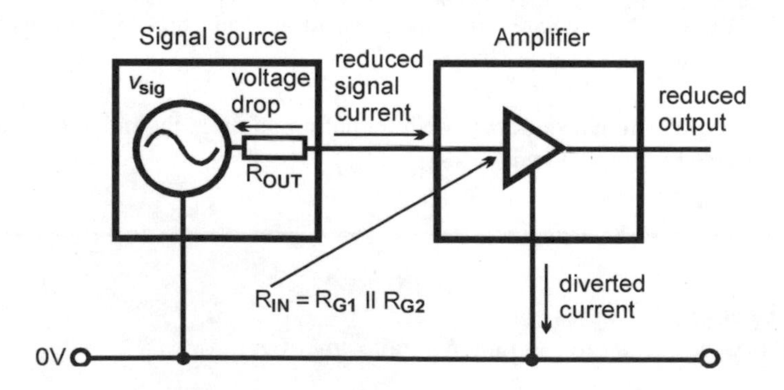

Figure 2.14 *Output and input resistance cause a partial reduction of signal amplitude.*

of the amplifier, there is a resistance (to ground and/or to positive) which diverts some of the signal instead of passing it all to be amplified. It is usually important to minimise this loss of signal.

As an example, the input resistance of the amplifier of Fig. 2.9 is equivalent to the two resistors in parallel. Given the values 390 kΩ and 560 kΩ, the input resistance is 230 kΩ. Note that we can use high-resistance biasing resistors because only negligible current flows out of the potential divider to the gate.

We can go further than this and connect the potential divider to the gate through another high-value resistor, as in Fig. 2.15. The 10 MΩ of the additional resistor is in series with the parallel resistance of the biasing resistors. The input resistance of the amplifier is the sum of these, totalling 10.23 MΩ. An amplifier with an input resistance as high as this takes very little current from any circuit feeding a signal to its input. However, only a

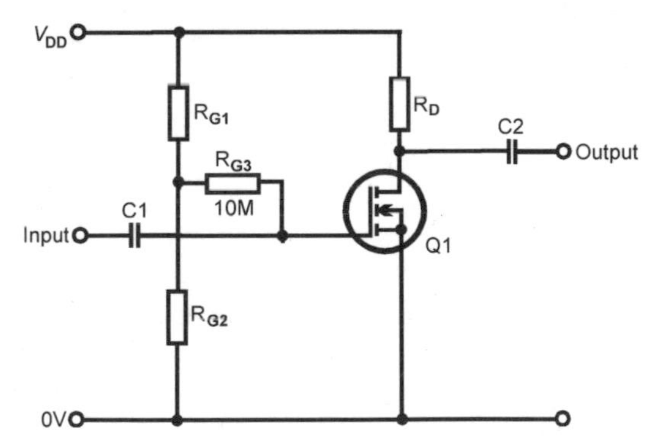

Figure 2.15 *Adding a third resistor to the biasing network further boosts input impedance, if needed.*

very small current flows to the gate of the transistor, so this kind of biasing network can be used only with FET transistors. The ease with which high input resistance is realised is one of the main advantages of MOSFET amplifiers.

Output resistance

Current drawn from the output of the amplifier flows through the drain resistor R_D from the positive rail. The output resistance of the amplifier is equal to the resistance of the drain resistor. This may vary from a few ohms to tens of kilohms in typical MOSFET amplifiers. In general, a low output resistance is desirable because this increases the ability of the amplifier to supply a strong signal (that is, relatively large currents) to following circuits. Note that, if a low output resistance is required, the drain resistor must be low. This means that a reasonably high drain current is required to generate a sufficiently high output voltage across it.

Current and power gain

The input current of a common-source MOSFET amplifier is so extremely small that, with any reasonable amount of output current, the amplifier has very high current gain. Because power = current × voltage, and output voltages are usually several millivolts or even volts, the power gain is also very high.

Source resistor

The performance of the common-source amplifier may be further improved by the addition of a source resistor (R_S, Fig. 2.16). The quiescent current through this raises the quiescent source voltage above zero and we must allow for this when biasing the gate. The action of the source resistor is to generate *negative feedback*. The way this works is as follows. An increased signal voltage leads to increased voltage at the gate. This increases the current through the transistor. Increased current through the source resistor leads to an increased source voltage. Raising the source voltage reduces the difference between this and the raised gate voltage. The gate-source voltage tends to remain unaltered.

Certain aspects of this are beneficial. If there are variations between the quiescent gate voltages of individual transistors due to variations in the value of parameters such as g_m, these tend to be compensated for by negative feedback. Also, if gate voltage varies with temperature or with the age of the transistor, these variations too tend to be opposed. Thus the source resistor gives greater stability and reproducibility to the circuit. The obvious disadvantage of negative feedback is that variations due to the signal are also cancelled out, leading to a considerable reduction in the gain

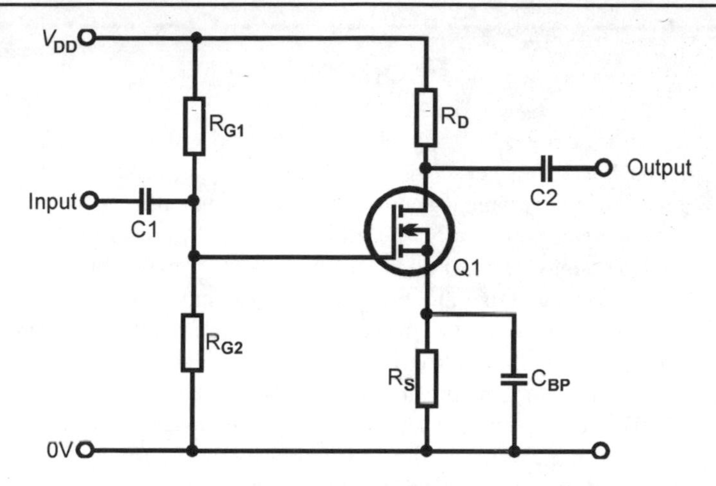

Figure 2.16 *A source resistor and a bypass capacitor further improve stability.*

of the amplifier. This effect is minimised by connecting a capacitor across the resistor, as in Fig. 2.16. This allows the signal to pass through and be lost to ground, instead of being fed back. For this reason we refer to this as a *by-pass capacitor*. Its action is to hold the source voltage steady against rapid fluctuations due to the signal. The signal is not affected by feedback so the gain of the amplifier is not reduced. Yet long-term variations, such as those due to temperature, are fed back and the amplifier is stabilised against these.

We can set the balance between stability and high gain by connecting the by-pass capacitor across part of the source resistor, as in Fig. 2.17.

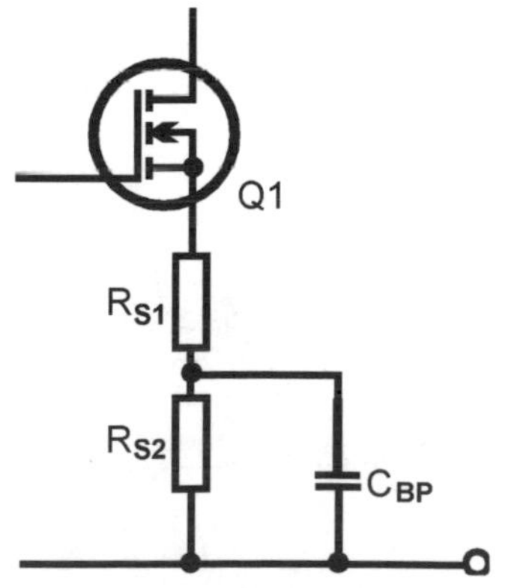

Figure 2.17 *Bypassing only part of the source resistance is sometimes used to obtain a compromise between stability and gain.*

Feedback

If a fraction of the output signal of a circuit or system is taken and routed back to an earlier stage, this is said to be *feedback*. Probably the most glaring example is acoustic feedback, when part of the sound from the loudspeaker of a public address system is picked up by the microphone, amplified and added to the original signal. This process continues, causing signal amplitude to increase until a loud whistling or screeching sound issues from the loudspeaker. This is *positive feedback* because the fraction of the output is *added* to the original signal. We obtain a similar effect by electrical feedback, in which we use a resistor or capacitor to add part of the output signal to the signal at an earlier stage. This is the basis of some oscillator circuits. In general, positive feedback leads to instability, either unintentionally or by design.

If part of the output signal is fed back in such a way as to be *subtracted* from the signal of the earlier stage, we have *negative feedback.* This leads to circuit stability, as explained by the example on this page. Several other examples of the use of negative feedback to improve circuit performance occur in later chapters.

Summing up

The three electrodes of an FET have names that describe what they do:

- Source - is the source of electrons flowing *into* the transistor.
- Drain - is the terminal through which electrons are drained *out of* the transistor.
- Gate - the electrode that controls *how many* electrons are allowed to flow through the transistor.

The relationship between input and output is given by: $i_D = g_m v_{IN}$, where g_m is the transconductance at any given value of i_D.

Adding the drain resistor converts i_D into an output voltage. $v_{OUT} = i_D R_D$.

The MOSFET common-source amplifier has many applications. Its main limitation is that is must be operated with small signals to avoid the effects of non-linearity (Figs. 2.5 and 2.6). Its advantages are:

- *High power gain.* This makes it suitable for power control circuits. In particular, VMOS transistors combine a high power rating with a very low 'ON' resistance (often 1 Ω or less). Two or more MOSFETs may be operated in parallel to share the current for very high power applications.

- *Rapid response time* compared with bipolar junction transistors, giving MOSFETs a very good frequency response and making them particularly suitable for radio-frequency amplifiers.

- *High input resistance*, making them ideal for amplifying signals from high-resistance sources (sensing applications, including medical sensing) and in instrumentation circuits.

- *Good temperature stability.* Typically, the gain of a MOSFET amplifier is reduced by only 0.2% per degree Celsius. Since the temperature coefficient is negative there is no thermal runaway, as with bipolar junction transistors (see Chapter 5).

Test Yourself

(Answers to numerical questions are in the Appendix)

1. Of what materials is a MOSFET made?

2. What is the distinguishing feature of the wiring of a common source amplifier?

3. What is meant by transconductance, and what is its unit?

4. Define the threshold voltage of a MOSFET and quote a typical range of values.

5. What name is given to the region of operation in which the drain current is relatively unaffected by the drain-source voltage but is affected strongly by the gate-source voltage?

6. Describe the function of the drain resistor in a common-source amplifier.

7. State the need for biasing the gate in a common-source amplifier. Calculate suitable standard values for the two biasing resistors, for a supply voltage of 12 V and a quiescent gate voltage of 3.9 V. What is the input resistance of the amplifier?

8. Why does an amplifier usually require an input coupling capacitor? In the circuit of question 7, what value capacitor gives a cut-off frequency of 200 Hz?

9. List the advantageous features of a common-source amplifier.

10. What are the functions of the source resistor and by-pass capacitor?

3
More FET amplifiers

In this chapter we look at another application of the n-channel enhancement type MOSFET, the source follower, and then consider amplifiers based on the junction field effect transistor (JFET).

MOSFET source follower amplifier

An alternative name for this is *common-drain amplifier,* because the transistor is in the *common-drain connection.* Input goes to the gate, output is taken at the source and the drain is common to both input and output circuits (Fig. 3.1). Like the common-source amplifier of Chapter 2, this amplifier has a pair of biasing resistors and is coupled by input and output capacitors. The difference is that there is no drain resistor and, for this reason, output is taken from the source of the transistor.

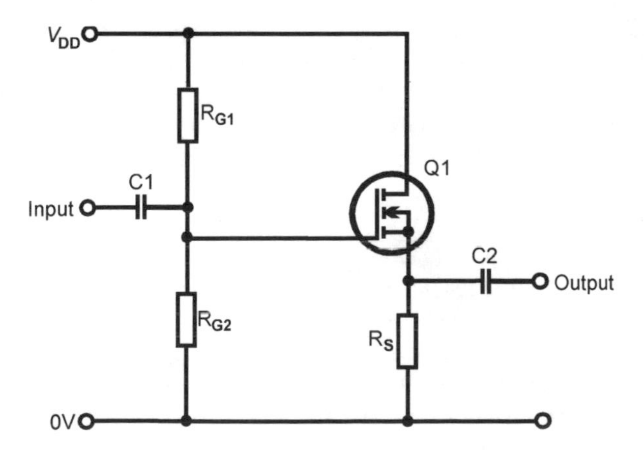

Figure 3.1 *In a common drain (source follower) amplifier, the drain is connected to the positive rail and the output is taken from the source.*

Given that the drain current at any instant is i_D, the output voltage v_{OUT} is given by:

$$v_{OUT} = i_D R_S$$

We know from the definition of transconductance that:

$$i_D = g_m v_{GS} = g_m(v_G - v_{out})$$

where v_G is the gate voltage. From these two equations we arrive at the solution:

$$v_{OUT} = \frac{R_S g_m}{1 + R_S g_m} \times v_G$$

The amplifier has a gain of less than 1. For values of R_S that are appreciably greater than $1/g_m$, this simplifies to:

$$v_{OUT} = v_G$$

The gain is only slightly less than 1 and so the output voltage is approximately equal to the gate voltage. This is the reason why this circuit is called a *source follower*.

Opposing current

The flow of current may be opposed or impeded by resistance, capacitance or inductance. Opposition due to capacitance or inductance is known as *reactance* (symbols, X_C, X_L). The total opposition due to any or all of three is known as *impedance* (symbol Z). All are measured in ohms. The effects of frequency on these are:

Resistance	No effect of frequency
Capacitative reactance	Decreases with frequency
Inductive reactance	Increases with frequency

The frequency must always be stated when quoting a value for reactance or impedance. Depending on the type of amplifier, the impedances are usually quoted for a standard frequency such as 1 kHz. This book is more concerned with the resistance component of impedance, so most references are to input and output resistance, but bearing in mind that reactance may play a significant role at low (X_C) and high (X_L) frequencies.

Input resistance

We have already seen in Chapter 2 that very little current is needed to provide a signal to a common-source amplifier. This is because the gate is highly insulated from the body of the transistor. The effect is enhanced because the high gate insulation means that high-value resistors can be used for biasing. The same is true of the source follower amplifier. Losses are low because both types of amplifier have high input *resistance*. But this is not the whole story. For completeness, we should state the input *impedance*, to take into account such effects as the capacitance between the gate electrode and the body of the transistor. Except at high frequencies (Chapter 8) the effects of reactance on input impedance can often be ignored.

Output resistance

We use the technique outlined in the box. Given unity gain, the open-circuit output voltage is equal to the input voltage, v_G (Fig. 3.2). The short-circuit output current is i_D and we already know that:

$$i_D = g_m v_G$$

We calculate output resistance as open-circuit voltage divided by short-circuit current:

$$\text{Output resistance} = v_G/i_D = v_G/g_m v_G = 1/g_m$$

The VN10AF transistor, for example has $g_m = 0.1$ S, so the output resistance is 10 Ω, which is reasonably low and ensures that the output signal is undistorted.

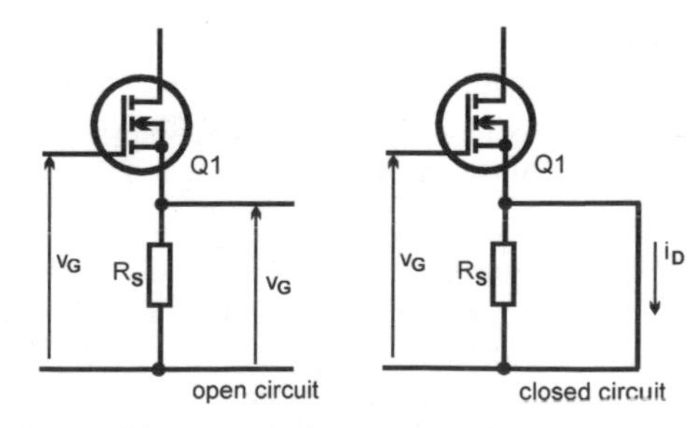

Figure 3.2 *Applying the open/closed circuit technique to evaluating the output resistance of a common-drain MOSFET amplifier.*

Evaluating output resistance

In Fig. 3.3 a source of a signal is represented by a voltage source v_s with its output resistance R in series with it (we will refer to resistance, but the argument is similar if there is an output impedance instead).

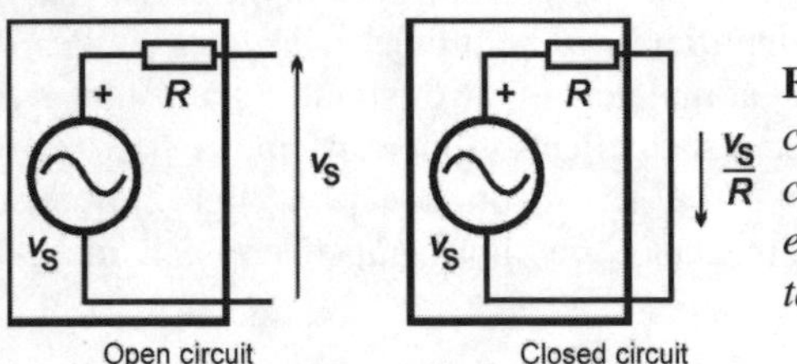

Figure 3.3 *Using open-circuit voltage and closed-circuit current for evaluating output resistance.*

To measure or calculate this we find:

The output voltage when the output is open-circuited. This equals v_s, because there is no current through the resistor, so there is no voltage-drop across it.

The output current when the output is short-circuited. This equals v_s/R.

$$\text{The output resistance} = \frac{\text{o/c voltage}}{\text{s/c current}} = v_2 \times \frac{R}{v_2} = R$$

This is a general technique for finding output resistance.

For comparison, the calculation of the output resistance of the common-source amplifier is illustrated in Fig. 3.4. Measuring voltages with respect to the positive supply rail, the open-circuit output voltage is $i_D R_D$. If the drain resistor is short-circuited to the positive supply rail, the current flowing through the short-circuit link is i_D. Again we calculate output resistance as open-circuit voltage divided by short-circuit current:

$$\text{Output resistance} = i_D.R_D/i_D = R_D$$

For the common-source amplifier, output resistance depends on the drain resistor. For the source follower, it depends on the transistor.

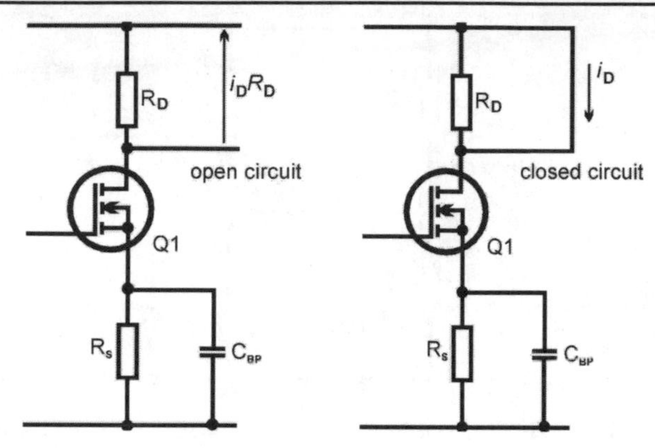

Figure 3.4 *The open/closed circuit technique shows that the output resistance of a common-source amplifier is equal to the resistance of the drain resistor.*

Applications

The high input resistance and low output resistance of the source follower, together with its unity gain, makes the source follower amplifier useful in the input stages of instrumentation circuits such as oscilloscopes or as a buffer connected to a high-resistance source such as a capacitor microphone. It is also useful in circuits such as sample-and-hold circuits, where a sampled signal has to be stored for a while on a capacitor while being measured or recorded.

Other types of MOSFET

The operation and properties of an amplifier based on a p-channel enhancement mode MOSFET are very similar to one using an n-channel MOSFET, except that polarities are reversed. The drain is held negative of the source, and the gate voltage is made negative of the source to produce a drain current. Fig. 3.5 shows a typical common-source circuit in which all the main features of the circuit based on an n-channel MOSFET are obvious. What is not quite so obvious in the symbol of the MOSFET is the direction of the small arrow-head on the internal link between source and substrate. Contrast Figs. 3.4 and 3.5 to note the directions. The main difference

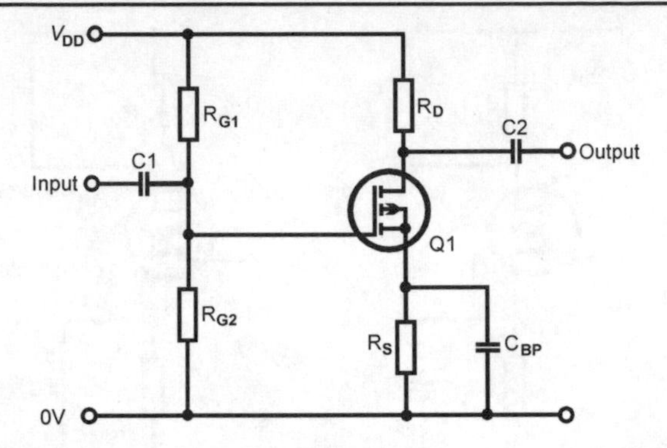

Figure 3.5 *This common source amplifier is based on a p-channel MOSFET. Note that V_{DD} is negative of the 0 V line.*

between amplifiers which use depletion MOSFETs, and those using enhancement types is the biasing required for the gate. In the n-channel type, the gate must be made negative of the source to turn the transistor off. Drain current increases as gate voltage is increased from its negative threshold, gradually becomes less negative, and passes through zero to become positive. The transfer characteristic is like that in Fig. 2.5 but is displaced to the left, in the negative direction.

Keeping up?

1 What is the approximate voltage gain of a source follower amplifier? What is the exact gain if the transistor has a transconductance of 0.8 S and a source resistor of 470 Ω?

2 What is the output resistance of the amplifier described in question 1?

3 List the types of MOSFET that are available.

Junction FETs

In MOSFETS, the gate is insulated from the body of the transistor by a thin layer of silicon oxide. In JFETs, as their name implies, the insulation is produced at a *junction*, the junction between n-type and p-type semiconductor. An n-channel JFET is formed from a block of n-type silicon into which a plug of p-type material (the gate) has been diffused (Fig. 3.6). The insulating layer is not an insulator as such, but a region of the n-type

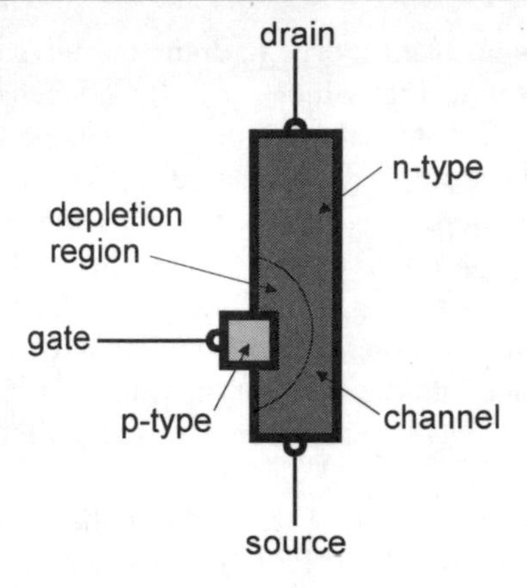

Figure 3.6 *Diagram of an n-channel field effect transistor. The depletion region occupies part of the width of the n-type material,leaving a relatively narrow channel for the conduction of charge from source to drain.*

material deprived of charge carriers. In other words, it is a *depletion region* formed when the pn junction is reverse-biased. This is the same as the depletion region formed in a reverse-biased diode. Because the p-type material in a JFET is more heavily doped than the n-type material, the depletion region is located mainly in the n-type material. To produce the depletion region, the gate of an n-channel JFET has to be held negative of the source. The most commonly used technique is to include a source resistor in the circuit, as in Fig. 3.7, so that the voltage generated across the resistor makes the source a few volts positive of the 0 V line. Only a single resistor is used to bias the gate, pulling it down very close to 0 V. In this way the gate is held a few volts negative of the source.

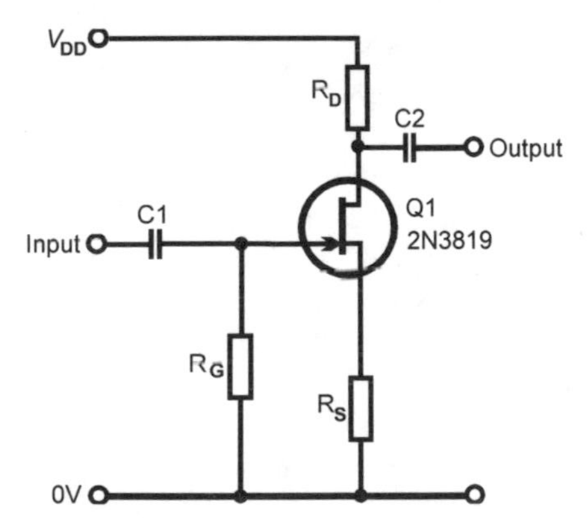

Figure 3.7 *In the basic common-source JFET amplifier, the source is a few volts positive of the 0 V line owing to the voltage developed across R_S. The gate is held negative of the source by the pull-down resistor R_G.*

Because the n-type material is continuous from source to drain, current can flow freely through the transistor. The ease with which it can flow depends on the width of the conductive channel, that is, the width of the region in which there is a sufficient supply of electrons to act as charge carriers. The more negative the gate, the wider the depletion region, the narrower the channel, and the smaller the current through the transistor. It is also possible to operate the transistor with positive v_{GS} up to about +0.5 V, but this is not often done. Above +0.5 V there is practically no depletion region and current begins to flow from the gate into the body of the transistor. So the JFET operates always in *depletion mode*. Enhancement mode JFETs are not possible.

The forward transfer characteristic of the JFET (Fig. 3.8) has the same upward-sweeping curve that is seen with MOSFETs, except that it is displaced to the left into the range of negative gate voltages. Transconductance increases with increasing i_D, resulting in distortion of large signals.

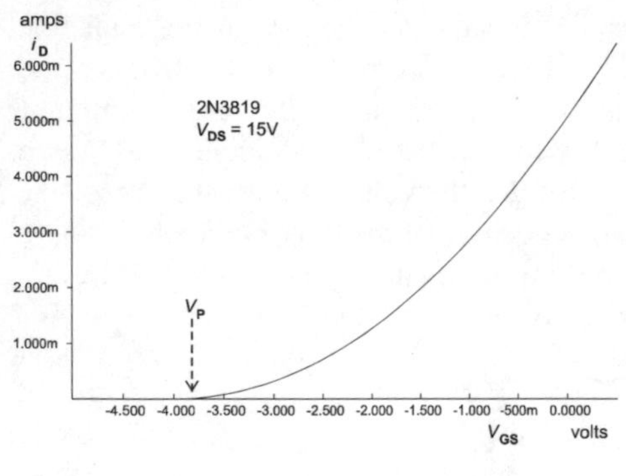

Figure 3.8 *The forward transfer characteristic of a JFET has the same upward sweeping form as that of a MOSFET (Fig. 2.5), but begins at a negative gate-source voltage and ends at about +0.5 V.*

The output characteristic of a JFET (Fig. 3.9) is also similar to that of a MOSFET except that v_{GS} ranges between about –8 V and 0 V. The point at which the depletion region extends fully across the transistor and the drain current is zero is sometimes known as the *pinch-off voltage*, V_P. This negative voltage is equivalent to the threshold voltage of MOSFETs.

The common-source n-channel JFET amplifier shares many features and characteristics with one based on a MOSFET. It includes drain and source resistors, and usually has input and output coupling capacitors and a by-pass capacitor. The main difference is in the biasing of the gate (Fig. 3.7). As explained above, this is held low by a *single* resistor, usually of the order of 1 MΩ, but its resistance can be made considerably higher if high input resistance is required.

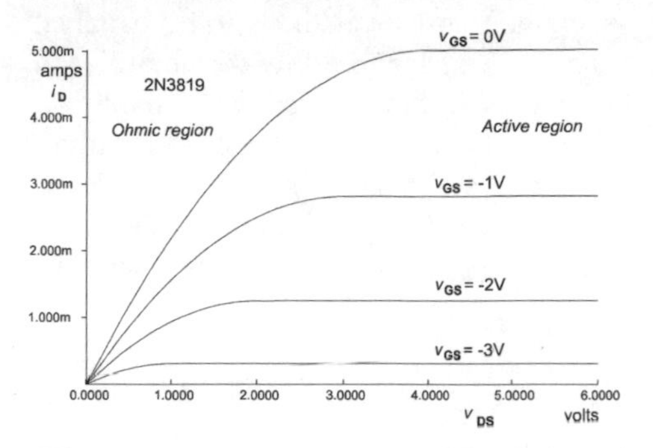

Figure 3.9 *The output characteristic of a JFET has features similar to those of a MOSFET (Fig. 2.6), except that the gate-source voltage is negative.*

When a common-source amplifier is based on a p-channel JFET, the drain is negative of the source and the gate is held positive of the source to reverse-bias the junction.

Source follower amplifier

This too has the same circuit as its MOSFET counterpart, except for the simplified biasing arrangement (Fig. 3.10). A disadvantage of a JFET source follower is that it has a higher output resistance. With a typical g_m of 2 mS, resistance is $1/g_m$, or 500 Ω. This can lead to distortion of the output signal.

Another disadvantage of the n-channel JFET source follower results from the fact that the gate voltage is held close to 0 V. If the supply voltage is 12 V, for example, the quiescent output voltage is ideally equal to half of this. Then the output signal is free to swing almost 6 V in each direction without distortion arising from saturation or bottoming out. But, if the gate is held

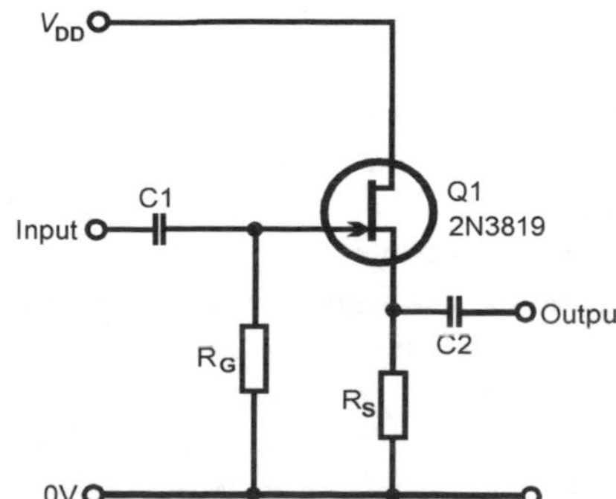

Figure 3.10 *A basic source-follower (common-drain) amplifier is limited to following signals of small amplitude, as shown in Fig. 3.11.*

6 V below the quiescent output level, the transistor is likely to be switched off for much of the cycle of the signal. We have to allow the amplifier to have a lower quiescent level and be content with a restricting the amplitude of the input signal. With a particular set of component values, the circuit of Fig. 3.10 has a quiescent output level of 2.8 V. This allows the amplifier to follow signals of small amplitude. Signals of larger amplitude, such as the 5 V input signal shown in Fig. 3.11, are not followed during their negative half-cycle.

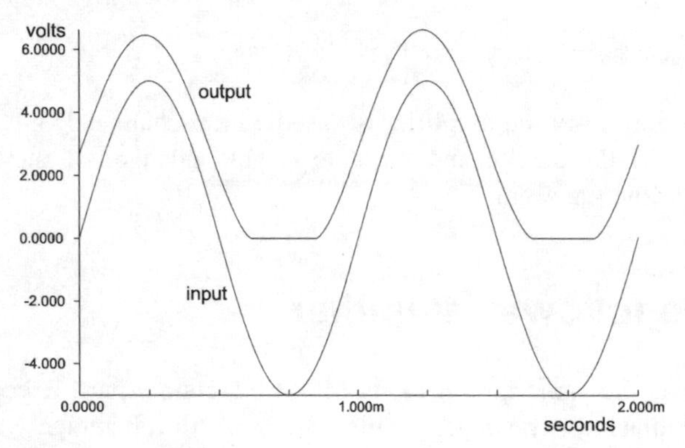

Figure 3.11 *A signal of 5 V amplitude (bottom curve) is fed to the amplifier of Fig. 3.10. But the quiescent output voltage is only about 2.8 V. This means that the output can not follow the large negative voltage swings.*

The circuit of Fig. 3.12 overcomes this problem. Here the gate bias resistor is tapped into the source resistor. Bias is provided by the quiescent voltage across R_{S1}. The voltage across R_{S2} raises all voltage levels, including that of the output by about 3.2 V, without otherwise affecting the operation of the amplifier. This allows the output to swing freely 5 V in either direction (Fig. 3.13).

It can be shown that the maximum power transfer between two circuits occurs when the power dissipation on the output side of one circuit is equal to that on the input side of the other. Thus, at most, a circuit is able to transfer only half its power to another circuit. Also, if maximum power is to be transferred, both the output circuit and the input circuit have to operate at the same power level, which is wasteful. In most amplification circuits, maximum *voltage* transfer is more important than maximum *power* transfer, since power can always be boosted from the battery or mains power supply to the amplifier. Thus the ideal transfer conditions are met when a circuit

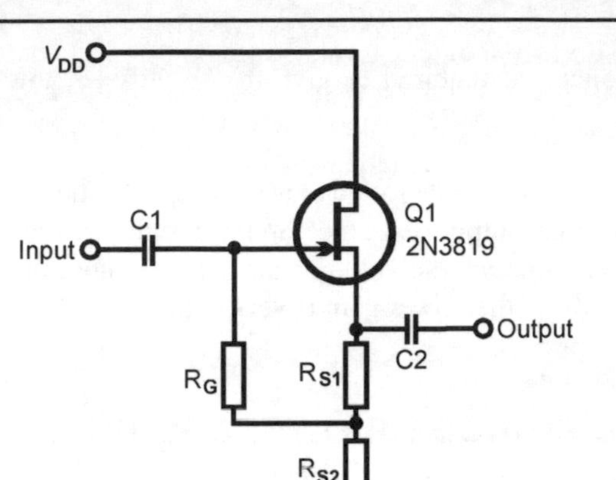

Figure 3.12 *Tapping the gate bias resistor into the source resistor raises voltage levels and allows the quiescent output voltage to sit midway between those of the power rails.*

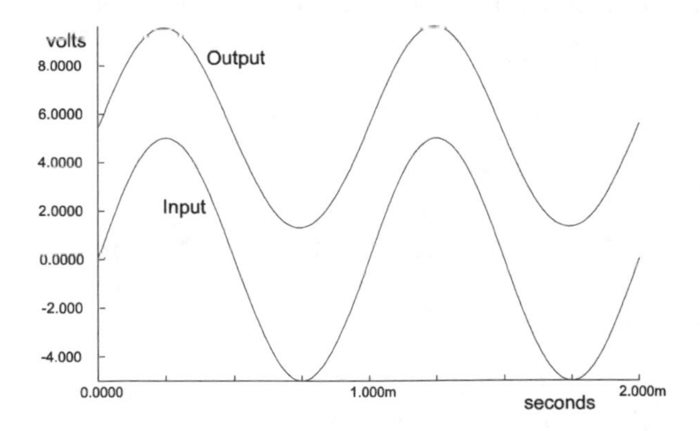

Figure 3.13 *The modified circuit of Fig. 3.12 has its output centred on 6 V (approx.), so that it can follow a swing of 5 V in either direction without distortion.*

with low output resistance feeds its signal to a circuit with high input resistance. With this arrangement, the receiving circuit does not overload the transmitting circuit and so degrade its voltage output. Conversely, the receiving circuit is fully responsive to the input signal. Because of their low input resistance, FET amplifiers can make good buffers between circuits that are not ideally matched. This is provided that the FET used has the high transconductance required to give the amplifier a low output resistance.

The primary use of the source follower is to act as a buffer between a circuit or device with high output resistance and a circuit with moderate or low input resistance. Although the voltage gain is never more than 1 (and often in the region of 0.9), the power gain is very high.

Improved source follower amplifier

In Fig. 3.14 the source resistor is in series with a constant current device. This consists of a second JFET (Q2) and a programming resistor, R_P. Transistors Q1 and Q2 are a matched pair, usually on the same chip. With a constant current flowing through Q2, an equal constant current must also flow through R_S, producing a constant voltage across it. Consequently, the output voltage follows the input voltage more closely than in the simpler type of source follower. In addition, the constant voltage across Q1 means

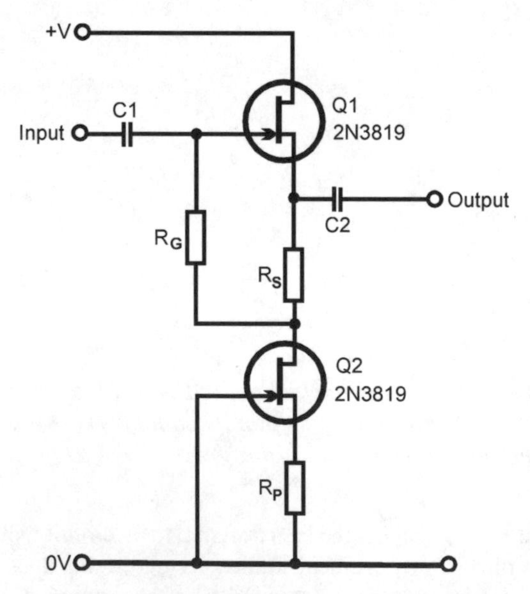

Figure 3.14 *In this version of the source follower amplifier, Q2 acts as a constant-current sink. This stabilises the current through Q1 and so makes the response more linear.*

that the voltage at the 'lower' end of R_G closely follows v_{in}. Only a small current (a few nanoamps) flows through R_G. This effect is as if R_G has a very much higher resistance than its actual resistance, so making the effective input resistance of the circuit equivalent to more than the actual value of the R_G.

Summing up

The MOSFET common-drain amplifier or source follower has a voltage gain of a little under unity. It has high input resistance and low output resistance, suiting it for linking stages with high output resistance to following stages especially those with low input resistance.

Junction FETs have the gate insulated from the body of the transistor by the depletion region at the pn junction. This means that, for an n-channel JFET, the gate terminal must be held negative of the source.

Junction FETs can be used to build common-source and common-drain amplifiers similar to those built from MOSFETs. In the common-drain (source follower) amplifier, a second JFET may be used as a constant-current device to improve its performance.

Test yourself

(Answers to numerical questions are in the Appendix)

1. Explain the difference between reactance and resistance.

2. Which terminal of the transistor is common to both input and output circuits in the source follower amplifier?

3. In a source follower, if the gate voltage is 3 V, the source resistor is 1 kΩ and the transconductance of the MOSFET is 0.7 S, what is the output voltage?

4. Write an expression for the output resistance of a voltage follower amplifier? What is the output resistance of a MOSFET with transconductance equal to 200 mS?

5. Describe or draw a diagram of the connections of a p-channel enhancement MOSFET in a source follower amplifier.

6. In what way does the gate of a JFET differ from that of a MOSFET?

7. In what ways does the biasing of the gate of an n-channel JFET differ from that of an n-channel enhancement MOSFET?

8. How can we design a JFET source follower so that its output is close to the mid-rail voltage?

9. Under what conditions is there maximum power transfer between circuits? What proportion of the power appears in the circuit to which it is transferred?

10. What are the best conditions for voltage transfer between two circuits?

4
Valve amplifiers

We are taking a break from solid-state electronics to look at an entirely different technology. Amplifiers based on valves (or *thermionic valves*, to give them their full title, or *vacuum tubes* to give them the name common in the USA) have been with us for far longer than transistor amplifiers. Yet, although one operates in the solid state and the other in a near vacuum, the principles of amplification and the circuitry involved are remarkably similar.

Diodes

The simplest form of valve, and the one which gives the devices their name, is the thermionic *diode*. Like the semiconductor diode, this has a one-way 'valve' action and is primarily used as a rectifier, either for producing DC power or as a detector in radio circuits. It does not have many applications in amplifiers but we describe it here as an introduction to valves in general. The diode, as its name implies, has two electrodes. These are enclosed in a glass envelope with a vacuum inside. In the simplest diodes (Fig. 4.1), the negative electrode or *cathode* consists of a filament of wire. The wire is heated by passing a current through it, causing a cloud of electrons to be concentrated around it. The production of electrons by heating gives rise to the description *thermionic* since, in the early days, the term 'ion' meant 'free particle'. In operation, the cathode is held at 0 V while the other electrode, the *plate* or *anode*, is raised to several hundred volts. The result of the electric field between cathode and anode is a current of electrons

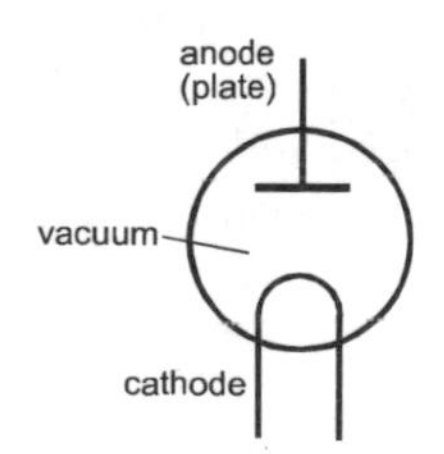

Figure 4.1 *A thermionic diode comprises two electrodes, anode and cathode, sealed in an evacuated glass tube.*

passing through the vacuum, from cathode to anode. Since only the cathode is heated, this is the only source of electrons and current flows only from cathode to anode. The one-way flow gives the device its name, which is applied to all devices based on this principle, even though their action is more complicated.

In most practical diodes, the filament is separate from the cathode. The filament is a fine wire wound on a former, situated on the axis of the vacuum tube. The cathode is a metal cylinder closely surrounding the filament. A low-voltage supply is used to heat the filament, which in turn heats the cathode. The high voltage supply is applied between the cathode and the anode. The anode too is cylindrical in shape, surrounding and co-axial with the cathode.

Field effect

The simplest amplifying thermionic valve has three electrodes, so it is called a *triode*. It has a cylindrical metal *grid* between the cathode and the anode (Fig. 4.2). As with the diode, the cathode is held at 0 V and the anode is several hundred volts positive. The grid is held negative of the cathode and in this way it corresponds to the gate of an n-channel JFET. As the electrons travel away from the cathode, under the influence of the electric field between cathode and anode, a proportion of them are made to return to the region of the cathode because of the oppositely-directed electric field between grid and cathode. The more negative the voltage on the grid, the greater proportion of electrons are returned and fewer reach the anode. This is how the current through the valve is controlled by small differences of

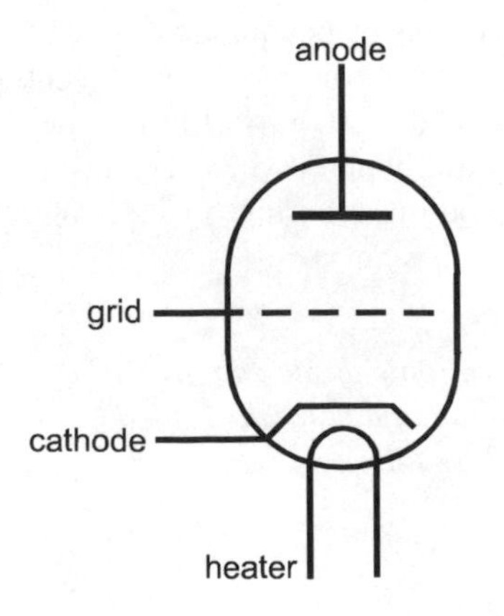

Figure 4.2 *The triode valve has a third electrode, the grid, between the cathode and anode.The charge on this controls the amount of current flowing through the valve. Some triodes use the heater as the cathode but the version shown here has separate cathode and heater.*

grid voltage. The behaviour of the triode parallels that of the JFET, and we can plot its forward transfer characteristic (Fig. 4.3). It looks very like the curve of Fig. 3.8. As in Fig. 3.8, we can measure the gradient at any point to obtain a value for the transconductance of the triode. Transconductances of valves are usually in the order of 5 mS to 12 mS, though they are more often expressed in the equivalent units, milliamps per volt.

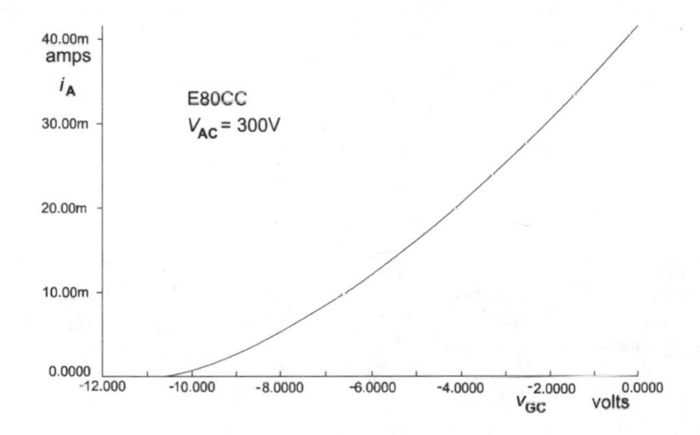

Figure 4.3 *The forward transfer characteristic of a triode has very much the same shape as that of a JFET.*

Common-cathode amplifier

This is the counterpart of the common-source JFET amplifier. There is a resistor (the cathode resistor R_K, Fig. 4.4) across which a voltage is generated to bias the grid in the negative direction. This corresponds to the source resistor in the JFET amplifier. A grid resistor R_G holds the grid close to 0 V and plays the same role as the gate resistor of the JFET amplifier. The input impedance at the grid is infinitely great. This is because no electrons are produced at the grid and so there is no way in which current can flow in or out through the grid terminal. The input resistance of the amplifier equals the resistance of R_G, which can be several megohms. The output resistance equals the resistance of the anode resistor, often referred to as the plate or load resistor.

In the way it works, including its input and output capacitors and its by-pass capacitor, the valve amplifier resembles its JFET counterpart, but there are a few important differences. The most obvious difference is that it requires a power supply of several hundred volts. This makes it less suitable for battery operation and considerably less portable than its solid state rival. On the other hand, it can handle signals of large voltage amplitude without distortion. The distortion of over-large signals is different from that in solid-state amplifiers, which usually show an essentially undistorted response up to a certain amplitude and a sudden clipping as maximum

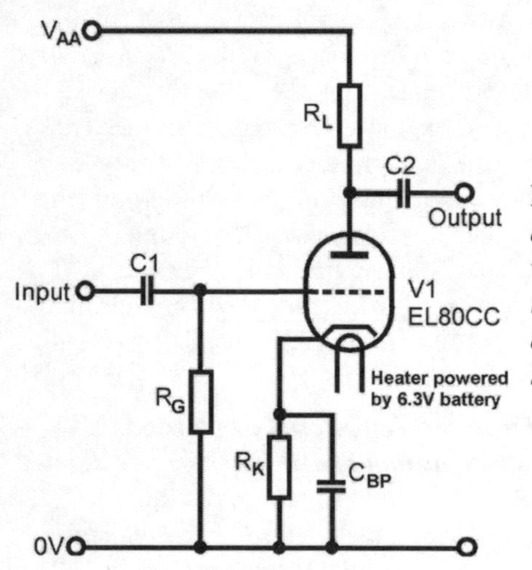

Figure 4.4 *A common-cathode amplifier based on a triode valve. Compare this figure with that of the JFET common-source amplifier (Fig. 3.7) and note the similarities.*

amplitude is approached. By contrast, distortion in a triode amplifier is more gradual. The signal does not *appear* to be severely clipped, yet there is a pronounced departure from the true sinusoidal shape.

There is one aspect of valve amplifiers which is still controversial. There are many who claim that valve amplifiers produce a truer and more pleasant sound than solid-state amplifiers. Whether or not this is correct or is an illusion is a subject of dispute but, certainly, many people are enthusiastic about valve amplifiers and this has led to a resurgence in the use of this technology in recent years.

Keeping up?

1. Name the electrodes present in a triode valve.
2. What kind of charge-carrier is found in a valve?

3. What part of a valve is equivalent to the gate of an FET?
4. Explain why current flows in only one direction through a valve.
5. How is the varying current through a valve converted to a voltage signal?

Tetrode valve

In the triode, there is capacitance between the cathode and anode. This is not important at low frequencies but becomes significant at high frequencies. It acts to damp out the signal. The remedy for this is to place a second metal grid in the valve (Fig. 4.5), between the original grid (now referred to as *the control grid*) and the anode. The second grid is known as the *screen grid*. It is close to the anode, and is held at about the same positive voltage, so that the electric field in the valve is maintained. Electrons flow from cathode to anode as before. But the screen grid is connected to the 0 V rail through a capacitor. For high-frequency signals, the screen behaves as if it is at 0 V and there is no capacitance effect. The tetrode therefore has much better high-frequency performance than the triode.

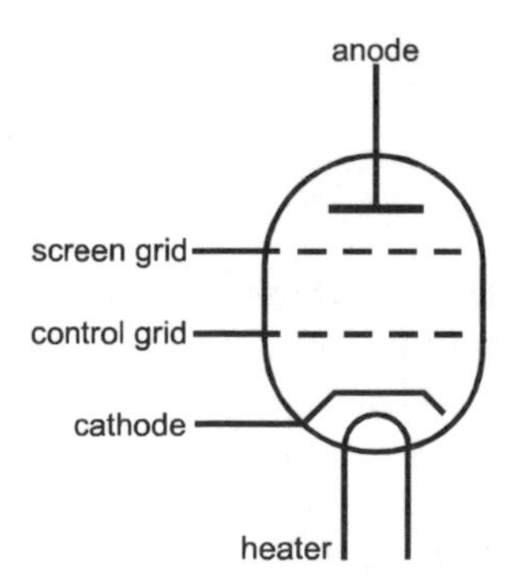

Figure 4.5 *The tetrode valve has a second grid, the screen grid to reduce the effects of capacitance between anode and cathode at high frequencies. This improves the frequency response of the valve. Flow of electrons between cathode and anode is controlled by the control grid.*

Fig. 4.6 is a common-cathode amplifier based on a tetrode. Input passes through C1 to the control grid, which is held a few volts above zero by the pull-down resistor R_G. The anode resistor R_L performs its usual function of current-to-voltage conversion and the signal passes through C2 to the output terminal. The screen grid is held close to the voltage of the anode by the pull-up resistor R_S, and is also grounded for high-frequency signals through C_S. The circuit is stabilised by the feedback from the cathode resistor R_K, with C_{BP} acting as bypass capacitor.

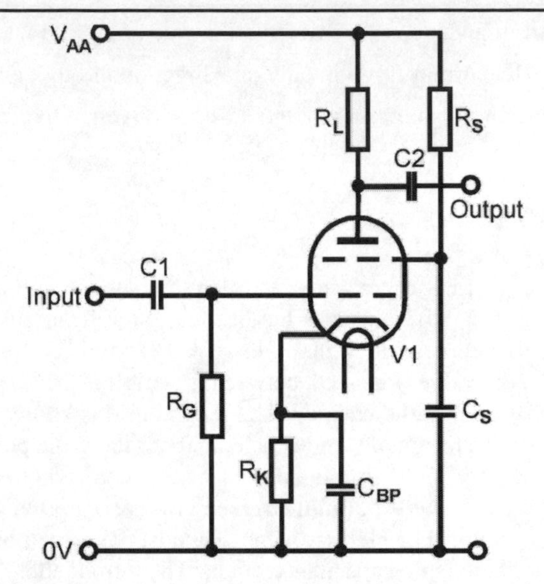

Figure 4.6 *In this tetrode-based amplifier the screen grid is held at high voltage by the pull-up resistor R_s. High frequency signals are by-passed to ground through C_s.*

Pentode valve

When electrons hit the anode of a tetrode, they cause the emission of electrons from the anode. These are known as *secondary electrons* and some of these are attracted to the screen grid. This is, in effect, a current passing in the reverse direction which distorts the signal. The pentode valve has a third grid to counter this (Fig. 4.7). The *suppressor grid* is situated between the screen grid and the anode, and is held at a low voltage. Its

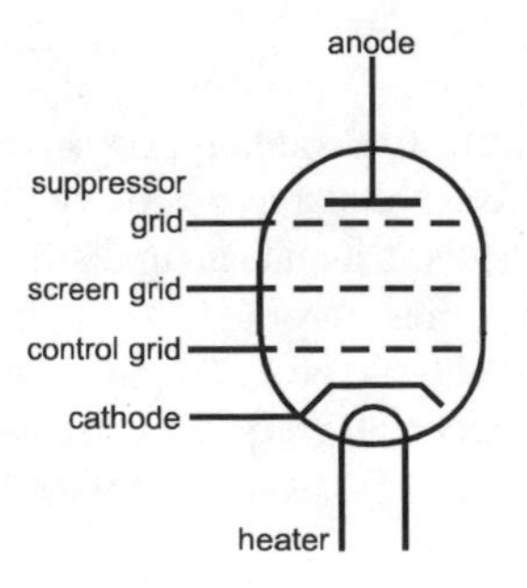

Figure 4.7 *The suppressor grid of a pentode valve is held at the same voltage as the cathode. Its function is to suppress the emission of secondary electrons from the anode.*

action is to repel the electrons emitted from the anode, sending them back toward the anode. The suppressor also exerts a repulsive force on electrons travelling from the cathode to the anode but these are more energetic and are able to pass through the meshes of the grid and reach the anode.

Fig. 4.8 is a common cathode amplifier based on a pentode. It has the same circuit as the tetrode amplifier (Fig. 4.6). In this circuit the suppressor grid is held at the same voltage as the cathode by an external connection. Valves may also be connected in other modes, such as common anode. The circuits behave in much the same way as their FET equivalents.

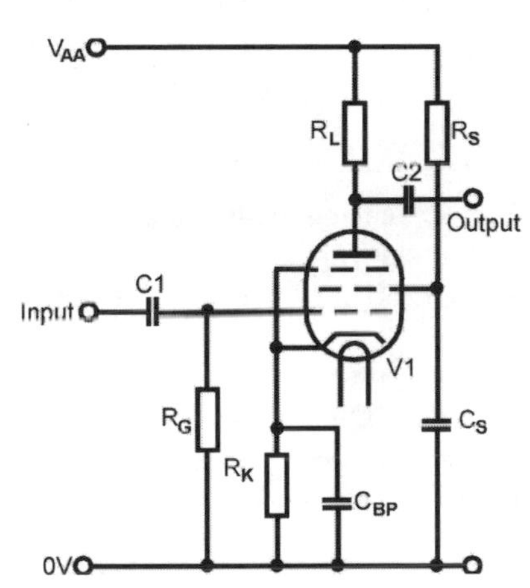

Figure 4.8 *In this amplifier, the suppressor grid of the pentode valve is connected directly to the cathode, so that both are at the same low voltage.*

Power valves

Compared with a transistor, a valve may seem to be a very powerful device. But, especially with high-power audio amplifiers, and radio transmitters, the two main present-day applications of valves, there is always a need for a device capable of delivering power rated in kilowatts rather than watts. Moderate increases in efficiency are effected by attention to the details of valve design. One popular form, the *beam power valve*, has special plates which produce a fan-shaped beam. This is broken into sheets by the control and screen grids. The well-defined beam leads to higher efficiency of operation and hence to higher power output. It also reduces the required screen current to only 5-10 % of the plate current instead of the 20-30 % of a conventional valve.

Medium power tubes dissipate between 5 kW and 25 kW, but in all tubes it is inevitable that some of the power is converted to heat, which must be removed if the tube is not to become damaged. For tubes rated at less than 50 kW we may employ an air-cooling system. A fan generates a strong current of air, which is directed around the valve by a system of baffles. For higher powers, from 50 kW up to 100 kW, it is better to employ water cooling, as in an automobile engine. Water is pumped around a system of pipes in thermal contact with the valve and then taken to a radiator. Water has a high specific heat capacity so it can absorb a lot of heat without reaching boiling point. But to convert the same amount of water to water vapour takes about 7 times as much energy. Vapour-phase cooling systems remove heat much more rapidly and are often used for valves rated at over 100 kW.

Summing up

A thermionic valve consists of a sealed, evacuated bulb containing a number of metal electrodes. Electrons are emitted from the heated cathode, and flow through the valve to the plate or anode, which is held positive of the cathode.

A diode is a valve with only a cathode and anode. Other valves have one or more grids in addition to the cathode and anode. These valves are named triodes, tetrodes, pentodes and so on, according to the number of electrodes they contain.

The grids are held at various potentials to regulate the flow of electrons. The most important grid is the control grid by means of which a varying input *voltage* can control the flow of *current* (electrons) through the valve.

In a typical valve amplifier the anode current flows through the anode resistor, generating an output voltage.

Valves are used for high-power amplification, and are also said by some to be superior for use in audio amplifiers.

Test Yourself

1. Why are thermionic valves so named?

2. Why can not current flow through a triode from the anode to cathode?

3. State a typical operating voltage for a valve.

4. Sketch the circuit of a simple triode common cathode amplifier.

5. What is the name and function of the extra grid present in a tetrode valve?

6. What is the name of the valve type which has five electrodes? What is the function of the suppressor grid and at what voltage is it normally operated?

5
BJT amplifiers

The *bipolar junction transistor* is the device we usually mean when we refer to 'a transistor'. The word 'transistor' is short for 'transfer resistor', as it is when applied to FETs. The word 'junction' refers to the fact that the action of the transistor depends on the pn junction, for the transistor contains two such junctions. The word 'bipolar' tells us that there are two kinds of charge carrier involved (electrons and holes) instead of only one as in the FETs, which are described as unipolar.

BJT action

The transistor is normally operated with its collector several volts positive of its emitter, the latter terminal often being connected directly to the 0 V rail. The base is usually made slightly positive of the emitter. Considering the base-emitter junction to start with, this is a forward-biased diode. As the base-emitter voltage V_{BE} is gradually increased from 0 V, it first has to overcome the virtual cell (0.6 V for a silicon transistor) before any current will flow from base to emitter. As soon as V_{BE} reaches about 0.6 V, current begins to flow. Looking at this in terms of the actual charge carriers involved, we have a flow of electrons entering the transistor by the emitter terminal and diffusing to the base-emitter junction (Fig. 5.3). There is also a flow of holes from the base terminal to the base-emitter junction. Electrons and holes combine at the junction.

But this is just part of the story. The base is only lightly doped so that relatively few holes are available. Only a small proportion (about 1 in a 100) of the electrons from the emitter are able to find a hole to combine with. About 99 % of the electrons remain uncombined. As we said earlier, the base layer is very thin. Electrons in this layer are very close to the depletion region between base and collector. They pass through the base layer and depletion region, and then come under the influence of the strong

BJT structure

There are two types of BJT, referred to by names which indicate their method of construction. An npn transistor (the other type is pnp) consists of a layer of p-type semiconductor sandwiched between two layers of n-type material (Fig. 5.1). The p-type layer is very thin, being much thinner than suggested in the figure. Electrical contact is made to all three layers, the base, the collector and the emitter. The

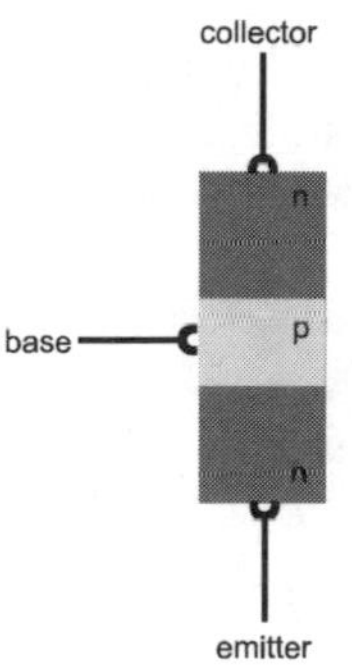

Figure 5.1 *An npn BJT consists of a thin layer of p-type material (not drawn to scale in this diagram) sandwiched between two layers of n-type material. Electrical connections are made to each layer.*

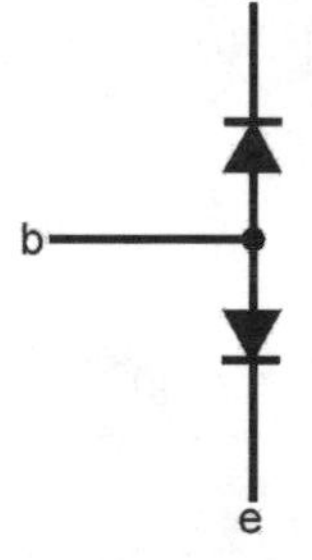

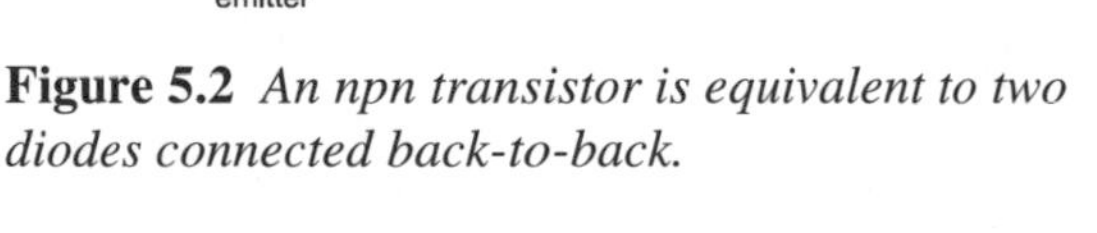

Figure 5.2 *An npn transistor is equivalent to two diodes connected back-to-back.*

two pn junctions occur on either side of the base layer. Inspection of the way the p and n layers are arranged shows that the transistor consists, in effect, of two diodes arranged back to back (Fig. 5.2). This being so, it is hard to imagine how it is possible for current to flow between the collector and the emitter. Current is made to flow by means of the BJT action.

electric field which exists between the base and the more positive collector. They are swept up by this field, flow through the n-type material and leave the transistor by the collector terminal. In summary, the emitter emits electrons, most of which are collected by the collector.

Although the base-emitter potential difference starts the electrons on their journey, the journey for 99% of them is completed by the action of the base-collector pd. What is important is that the number of electrons recombining with holes at the base-emitter junction is a fairly fixed proportion (1 %) of the total. Conversely the number that pass through to the collector is also a fixed proportion (99 %). In terms of conventional current, the current I_C flowing in through the collector terminal is about 100 times the current I_B flowing in through the base. Fig. 5.4 shows how we measure these currents. Because only 1 % of electrons recombine:

$$\text{Current gain} = \frac{I_C}{I_B} = \frac{99}{1} \approx 100$$

Gain varies from one transistor type to another but is usually in this order. The point of comparison with FETs and valves is that both input and output are currents. The BJT is a *current amplifier.*

We also note that the current flowing out through the emitter terminal is the sum of base and collector currents:

$$I_E = I_B + I_C$$

Since I_C is so much larger than I_B:

$$I_E \approx I_C$$

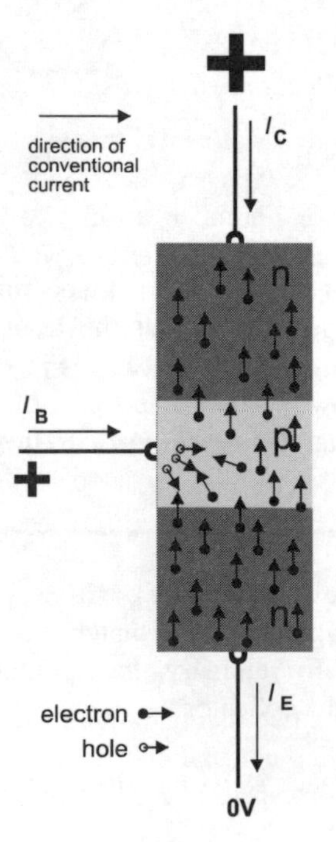

Figure 5.3 *The flow of electrons and holes in an npn transistor. In the base layer, a few of the electrons from the emitter combine with holes from the base, but the majority of the electrons flow through to the collector.*

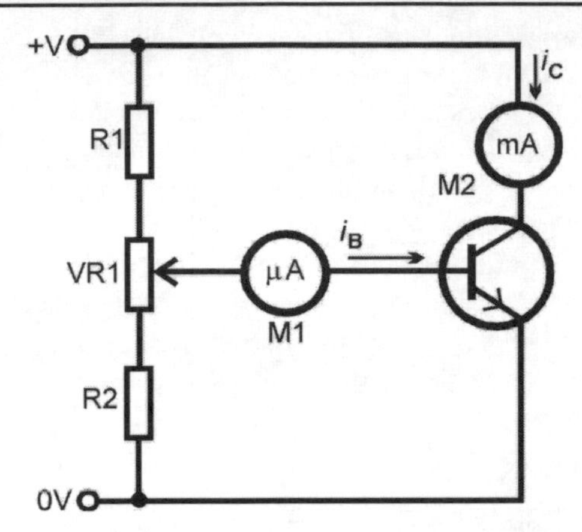

Figure 5.4 *This circuit is used for measuring the forward transfer characteristic of the BJT. Note that the base current is usually measured in microamps, but the collector current is larger and is usually measured in milliamps.*

Forward transfer characteristic

To examine the currents in a particular transistor we plot its *forward transfer characteristic*, using a circuit of the type shown in Fig. 5.4. We vary VR1 and record the output current I_c for a number of input currents I_B. The result is similar to the plots we have made for other amplifying devices (Figs. 2.5, 3.8 and 4.3), except that this plot has currents along *both* axes (Fig. 5.5). The line is straight in its 'working region' but there are regions at low and high current, where the curve slopes differently.

One way of defining gain is to take a point on the working region of this curve and read off the base current and collector current at that point. This tells us what collector current we obtain when the base current has a given value. Then we calculate the gain from:

$$\text{gain} = \frac{\text{output}}{\text{input}} = \frac{I_C}{I_B}$$

Note that we are calculating a ratio between two identical quantities, so the result is a pure number with no units. This calculation depends on the amount of collector current flowing at *a given value* of I_B. It is known as the

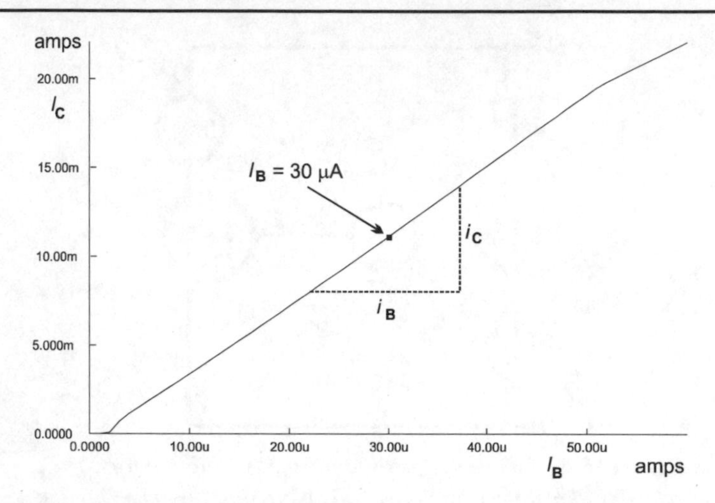

Figure 5.5 *The forward transfer characteristic of a BC548BP bipolar junction transistor. To find h_{fe} when I_B = 30 μA, measure i_B and i_C.*

large signal current gain, and has the symbol h_{FE}.

Another way of defining gain is to measure the slope or *gradient* of the curve at a particular point. In Fig. 5.5 we measure i_B, a small *change* in base current, and i_C, the corresponding small *change* in collector current. Dividing one by the other we obtain the *small signal current gain*, h_{fe}:

$$h_{fe} = \frac{i_C}{i_B}$$

Again we have a *ratio*, with no units. Because the curve of Fig. 5.5 is more-or-less a straight line, h_{fe} is almost the same thing as h_{FE} and we often do not bother to distinguish between them. When we refer to current gain, without saying which one, we usually mean h_{fe}. This also has the symbol β.

Output characteristic

This is measured in the same way as the output characteristics of FETs and valves (Figs. 2.6, and 3.9). The curves (Fig. 5.6) rise very steeply at low voltages (the *saturation region*) and then turn over and run almost horizontally (the *linear region*). Note that the curves are much more evenly spaced in this region than they are in the corresponding region of an FET. This

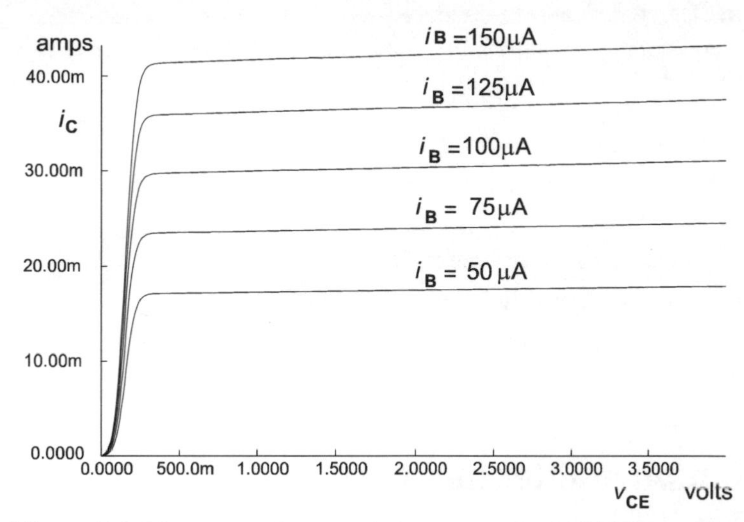

Figure 5.6 *The output characteristic curves of a BC548BP transistor.*

reflects the fact that the plot of the forward transfer characteristic of a BJT is almost a straight line, whereas that of an FET curves upward (Fig. 2.5). As a result of this a BJT is able to amplify a relatively large signal without distortion.

Looking more closely at Fig. 5.6 we see that the curves slope slightly upward in the linear region. This shows more clearly if we extend the range over which voltage is plotted. Then we can see that the linear regions of the curves do not have equal slopes. If the linear parts of the curves are projected back towards the left in the figure, they converge on to a point on the *x*-axis. This is known as the *Early point* and lies at approximately –100 V. This is a result of the *Early effect*. What happens is that an increase in V_{CE} causes an increase in V_{CB}. In other words, the collector-base junction becomes more strongly reverse-biased. This increases the width of the depletion region. The base is very thin and any increase in the depletion region makes it appreciably thinner. There is less chance of holes and electrons in the base region recombining, leaving slightly more electrons available to flow on toward the collector. This leads to a slight increase in collector current and thus to a curve that slopes upward with increasing V_{CE}. For this reason the performance of a BJT amplifier is more dependent on power supply level than it is with an FET amplifier, though the effect is not large.

Keeping up?

(All 'transistors' referred to in these questions are bipolar junction transistors)

1. Why is the transistor described as 'bipolar'?
2. At what base-emitter voltage does collector current begin to flow through a transistor?
3. What is the typical current gain of a transistor?
4. What quantity is represented by h_{fe}? What are its units?
5. In what region of the output characteristic curve do we usually operate the transistor?

Common emitter amplifier

The circuit of a basic CE amplifier (Fig. 5.7) has mostly the same features as the common source FET amplifier (Fig. 2.15). An important difference is that an appreciable current flows into the base of the transistor. Therefore we can not use high-value resistors to bias the base. If we are aiming for a quiescent collector current of 1 mA, for example, the base current will be about one hundredth of this, or 10 μA. If the base current is to be of this size, the current flowing through the potential divider must be at least 10 times this amount, or 100 μA. Fig. 2.10 explains the need for this. Given a supply voltage of 9 V, the total of the resistor chain must be no more than 90 kΩ. A consequence of this restriction is that the input impedance of a common emitter amplifier can never be as high as that of a common source amplifier.

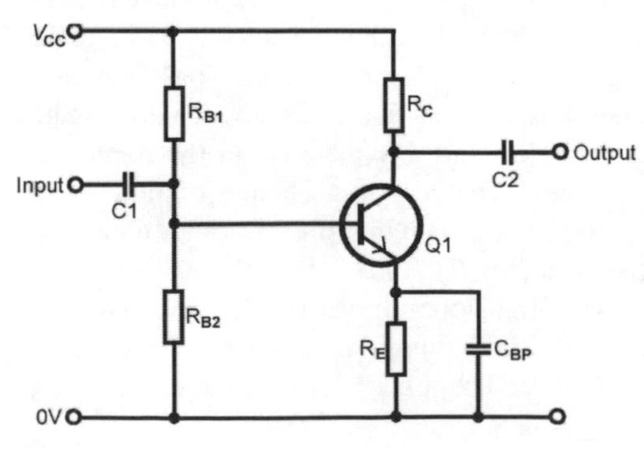

Figure 5.7 *Compare this common-emitter BJT amplifier with the common-source FET amplifier of Fig. 2.15.*

A related point is that base current flows into the base terminal and out at the emitter terminal, passing through semiconductor material on the way. This offers resistance to the flow of current and can be thought of as an additional resistor inside the transistor, as in Fig. 5.8. We refer to this as the *emitter resistance,* r_e. For a typical collector current of 1 mA, this has a value of about 25 Ω and must not be confused with the resistance of the emitter resistor R_E, which may or may not be present. Normally r_e may be ignored because it is in series with R_E which has a much larger resistance.

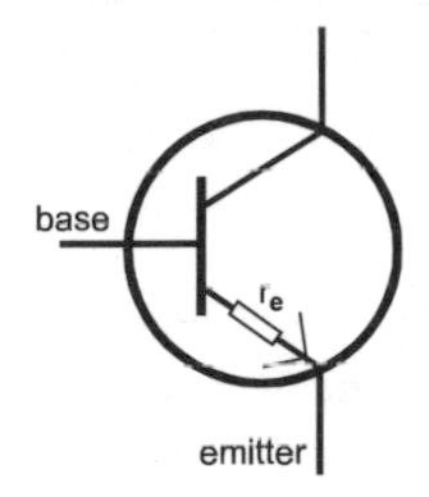

Figure 5.8 *The semiconductor material of the BJT offers resistance to the flow of current. The flow from base to emitter encounters the emitter resistance of about 25 ohms. In the figure, it is represented by a resistor symbol drawn inside*

Now to calculate the voltage gain of the amplifier of Fig. 5.7, with R_E present. If the input changes by a small amount v_{in}, this is passed across the input capacitor C1 and becomes a change in the base voltage v_b. Because of the base-emitter junction, the potential difference between base and emitter remains more-or-less constant, so there is a change, v_e in the voltage at the emitter and:

$$v_e = v_b = v_{in}$$

The change in v_e produces a change i_e in the current flowing through R_E:

$$i_e = \frac{v_e}{R_E}$$

But we have said that emitter current is almost equal to collector current, so changes in these are equal:

$$i_c = i_e$$

Finally the change in output voltage v_{out} is:

$$v_{out} = -R_C i_c$$

The negative sign indicates that an increase in i_c causes a drop in v_{out}. This is an *inverting amplifier*. Putting all these equations together:

$$v_{out} = -R_C i_c = -R_C i_e = -\frac{R_C v_e}{R_E} = -\frac{R_C v_b}{R_E} = -\frac{R_C v_{in}}{R_E}$$

From this we obtain:

$$\text{voltage gain} = \frac{v_{\text{out}}}{v_{\text{in}}} = -\frac{R_C}{R_E}$$

The interesting thing about this result is that the voltage gain depends only on the values of the collector and emitter resistors. It is not affected by the current gain of the transistor.

Biasing

Biasing by means of a potential divider produces a steady voltage at the base and allows for the greatest stability. Having decided on the quiescent current we calculate the emitter voltage, add 0.6 V to this to bring the transistor into conduction, and calculate the resistor values required for the potential divider.

A simpler means of biasing is to have only one resistor, as in Fig. 5.9. In an FET amplifier the minutely small current flowing to the gate produces a negligible voltage drop across R_G, so that the gate is held close to the positive power rail. We adjust the biasing by setting the source resistor. But in a BJT there is appreciable base current and an appreciable voltage drop across the biasing resistor. If we know the required base voltage, we can

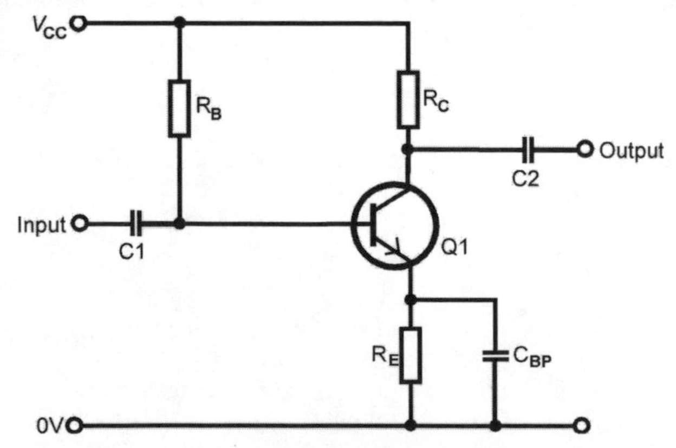

Figure 5.9 *This common-emitter BJT amplifier uses only a single biasing resistor. It is less stable than the amplifier of Fig. 5.7.*

calculate a suitable value for the resistor. For example, if the quiescent emitter voltage is 1 V, the quiescent base voltage is 1.6 V. Given a 6 V supply voltage, the required drop is 4.4 V. If the required base current is 10 μA, the resistor must be R_B=4.4/0.00001 = 440 kΩ.

This is a simple circuit, using only one resistor instead of two, but it has a big disadvantage. The calculation assumes that we know what base current is required to produce a given quiescent collector current. To know this we need to know h_{fe}. Unfortunately h_{fe} varies widely between different transistors of the same type, due to differences arising during manufacture. It may easily happen that the actual h_{fe} is 160 instead of the assumed 100. The base current is only 6.25 μA and the resistor value should be 704 kΩ. Another factor is that h_{fe} varies with temperature, so amplifiers built in this way are not temperature stable.

The circuit with a single biasing transistor is improved by connecting one end of the resistor to the collector terminal instead of to the positive supply line (Fig. 5.10). At the same time we halve its value. This circuit obtains its stability because of feedback. If h_{fe} is greater than was assumed in the calculation (either because the particular transistor used has an especially high value or because of temperature effects), the quiescent i_c is *increased*. This causes a fall in the collector voltage. The voltage drop across R_B is

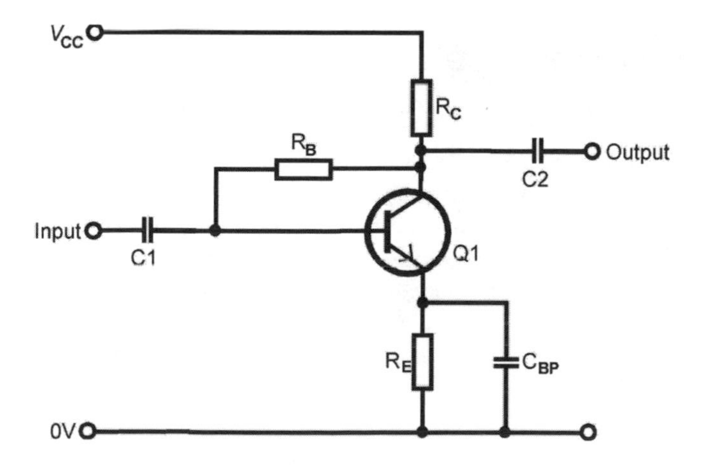

Figure 5.10 *The stability of the amplifier of Fig. 5.9 is improved by connecting the biasing resistor between the collector and the base and halving its value.*

reduced, resulting in a reduction in base current. This tends to *reduce* i_c. Feedback is *negative* and the value of i_c is reasonably stable. The reverse occurs if i_c is smaller than assumed.

Input resistance

The input resistance is primarily due to the two (or one) biasing resistors connecting the input point to the power rails. As we have said, this is lower than for FET amplifiers because the biasing resistors are necessarily of lower value. There is also a third resistance, because in the BJT we have a base current flowing through the transistor and the emitter resistor to the 0 V rail. Strictly speaking we should include the emitter resistance r_e in series with this but, since R_E is often of the order of 1 kΩ, we can usually (but not always) ignore the 25 Ω due to r_e. Although the value of R_E may be only 1 kΩ, this is not the value which contributes to the input resistance. This is because I_B is only about one hundredth of the current flowing through R_E. The effect is as if R_E is one hundred times bigger than it really is. If R_E is 1 kΩ and h_{fe} is 100, the effective value of R_E is 100 kΩ. In parallel with bias resistors of typical values, this resistance may usually be ignored.

Output resistance

On the output side we have the collector resistor in parallel with the resistance of the collector. The latter is equivalent to several megohms, so the only significant resistance is the collector resistance. In an amplifier designed to operate with a quiescent current of 1 mA, and assuming a 9 V supply, the collector resistor needed to bring the quiescent output voltage to half the supply is 4.5 kΩ, and this will be the output resistance of the circuit. An amplifier intended to operate with 100 mA quiescent current on the same supply would have a collector resistance (and therefore output resistance) of only 45 Ω.

In discussing input and output resistance, we have overlooked the matter of capacitance and its frequency-dependent effects. Together with the effects of resistance, these contribute to the total input and output *impedances* of the amplifier. These effects are important only at high frequencies so we shall examine them in Chapter 8.

Temperature effects

BJTs, especially those based on germanium, are more subject than FETs to the effects of temperature. This is because the base-collector junction is reverse-biased and there is a leakage current through it. The size of the leakage current is proportional to temperature. The leakage current is small but has the same effect as a current supplied to the base from an external source. The current is amplified by the usual transistor action, so giving rise to a collector current about 100 times bigger. So any increase in temperature results in an increase in the leakage current which in turn leads to an increase a hundred times greater in collector current. With a relatively small increase in temperature the collector current may easily become increased by amounts that approach the magnitude of the normal signal currents. This results in degradation of the signal.

A more serious effect may occur if temperature rises even higher. Then there is such an increase in collector current that the transistor begins to overheat. This brings about an even greater increase in temperature which raises the collector current even more. The effect is cumulative, collector current increasing again and leading to further temperature rise. We have *thermal runaway*, which usually results in destruction of the transistor. This problem, absent from FETs, is most serious with germanium BJTs but can also occur with silicon BJTs. It may be avoided by ensuring that the transistor is attached to a heat-sink adequate to limit temperature to a safe level. It is also possible to design the circuit so that collector current itself can not rise above a suitable level.

Transconductance

When we are discussing FETs, the transconductance is the only available indicator of the amplifying capability of the transistor. Transconductance is a current divided by a voltage, so its unit is the siemens. With BJTs we measure both input and output in the same unit, the ampere. The ratio between them, h_{fe}, is the most convenient way of indicating amplifying ability. *Current gain*, being a current divided by a current, has no unit.

The concept of transconductance can be applied to BJTs too, and produces an interesting result. As in the FET, the quantities concerned in the transconductance of a BJT are the input voltage and the output current. The input voltage is the voltage across the base-emitter junction, v_{BE}. The output current is the collector current, i_C. The transconductance of the BJT is given by:

$$g_m = i_C/v_{BE}$$

This equation defines transconductance for a BJT. It can also be shown, to a reasonable degree of approximation, that:

$$g_m = 40I_C$$

where I_C is the quiescent collector current, expressed in milliamps and g_m is in milliamps per volt. This equation reveals that the transconductance depends not on the individual transistor but on the collector current that we pass through it. This current depends on the component values and connections of the circuit in which the transistor is operating. Contrast this with the FET in which transconductance varies from transistor to transistor and depends on manufacturing differences. The variation of the transconductance of a BJT with collector current is a source of distortion when signals are large.

Summing up

There are two types of bipolar junction transistor, described as npn and pnp.

Both consist of three semiconductor layers, the collector, the base and the emitter.

In a suitably connected npn transistor, a small current flowing into the base and out through the emitter results in a much larger current flowing in through the collector and out through the emitter.

The current gain of the transistor is defined as the collector current divided by the base current, and is in the region of 100. The transconductance of a BJT depends only on collector current.

The common emitter amplifier has a resistor in its collector circuit. The collector current through this produces the voltage output of the amplifier. The value of the resistor is chosen so that, when the quiescent current is flowing, the output voltage lies approximately midway between that of the two supply lines.

There may also be an emitter resistor, in which case the voltage gain is $-R_C/R_E$. In the absence of the emitter resistor the voltage gain is not infinite because there is an internal resistance r_e due to the material of the emitter region. This is typically 25 Ω when the emitter current is 1 mA.

The base of the transistor is biased using two resistors as a potential divider, or by a single voltage-dropping resistor wired either to the positive rail or to the collector. The first arrangement is the most stable.

The input resistance of the common-emitter amplifier equals that of the biasing resistors in parallel with each other and with the emitter resistor multiplied by the gain. The total input resistance is generally lower than that of the equivalent FET common-source amplifier. The output resistance equals the resistance of the collector resistor.

Amplifiers based on BJTs are subject to thermal runaway. They may be protected by limiting the collector current and by mounting the transistor on a heat sink.

Test Yourself

1. Describe the structure of an npn transistor.

2. Why is the collector current of a BJT amplifier much greater than the base current?

3. Which type of amplifier typically has the higher input resistance, common-source or common-emitter?

4. What is the typical value of the emitter resistance, r_e?

5. What factors determine the voltage gain of a common-emitter amplifier?

6. Calculate the input resistance of a common-emitter amplifier with an emitter resistor of 470 Ω, if h_{fe} equals 125.

7. List the three ways of biasing a common-emitter amplifier and state which is the most stable.

6
Other BJT amplifiers

The common-emitter amplifier is perhaps the most frequently used of transistor amplifiers, but there are other kinds of BJT amplifier, each with its special features and applications. In this chapter we look at these and discuss their useful properties.

Common-collector amplifier

A common-collector amplifier is the BJT equivalent of the FET common-drain amplifier and is used for similar purposes. This is why its alternative name is the *emitter-follower*. Fig. 6.1 depicts a typical common-collector circuit. As usual, capacitors couple the circuit to preceding and following stages. But these capacitors are not an essential part of the amplifier. If input or output quiescent voltage levels are suitable, the amplifier may be wired directly to adjacent stages.

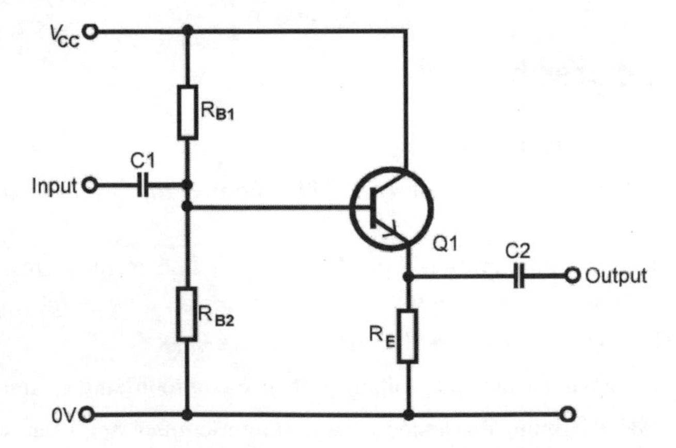

Figure 6.1 *A common-collector BJT amplifier is also known as an emitter-follower, because the output voltage at the emitter is always close to the input voltage.*

R_{B1} and R_{B2} are used to bias the transistor into conduction. Normally the base current is such as to produce a collector current which generates a voltage across R_E equal to half the supply voltage. This makes the quiescent output voltage sit nicely between the two extremes and allow the output signal to be as large as possible without risking its being clipped.

The base-emitter junction is forward biased so the base current flows through the junction and becomes part of the emitter current. It flows first through the internal emitter resistance (r_e, Fig. 5.8) then through the emitter resistor R_E. Together these act as a potential divider, so that the output voltage is less than the input voltage. However, r_e is small (typically 25 Ω) and R_E is large (several kilohms), so the output voltage is generally 99 % or more of the input voltage. This is close enough to unity gain. Since output follows input very closely, the output is in phase with the input. In other words, this is a *non-inverting amplifier*.

The input resistance of the amplifier is equal to the resistance of the biasing resistor (or resistors) in parallel with the load resistance (which includes R_E). The biasing resistors typically have values rated in tens of kilohms. In the case of the load resistance, looking into the base of the transistor we see R_E multiplied by the gain h_{fe} , so R_E *appears to be* several kilohms multiplied by 100 or more, that is, several hundred kilohms. As a rule, the amplifier is feeding its signal to a circuit connected in parallel with R_E. The input resistance of the amplifier thus comes to be the parallel resistance of:

- the biasing resistors,
- R_E multiplied by the current gain, and
- the input resistance of the load circuit.

Usually the resulting input resistance is high, so that we can think of a typical emitter follower amplifier as having a high input resistance.

By contrast, the output resistance is approximately equal to r_e plus the output resistance of the source circuit (for example a signal generator or microphone) *divided by* the gain of the transistor. If the source has low output resistance, we may ignore it and consider only r_e, which is only a few tens of ohms. If the source has very high output resistance we may take R_E to be wired in parallel with this, so reducing the output resistance of the amplifier. In most cases the result is a very low output resistance.

With its high input resistance and low output resistance, the common-collector amplifier is particularly useful as an impedance-matching (or resistance-matching) stage. That is to say, if we have a circuit or device (such as a sensor, for example) which has low output resistance and connect it to a monitoring circuit with low input resistance, we will probably find that the signal passing from one to the other is markedly reduced in amplitude. It may be almost lost. But if we put an emitter follower between the two, the follower draws very little current from the sensor, so maintaining the amplitude of its signal. At the same time the follower provides ample current to drive the monitoring circuit.

Bootstrapping

The concept of pulling oneself up by one's own bootstraps is a comical one but is a lucid description of the function of capacitor C3 in Fig. 6.2. Omitting R4 and C3, we have a typical common-collector amplifier. Its output follows changes in the input but is usually six or seven hundred millivolts lower. Its input resistance, given the typical values quoted in the caption of Fig. 6.2, is the resistance of R1 and R2 in parallel. On a simulator, this is equal to 8.3 kΩ. If we then insert R4 and connect C1 to the base side of R4, the input resistance becomes 17.9 kΩ, the increase being due to the 10 kΩ of R4 now in series with the paralleled resistances of R1 and R2. Apart from this, the action of the amplifier is unaffected. The final

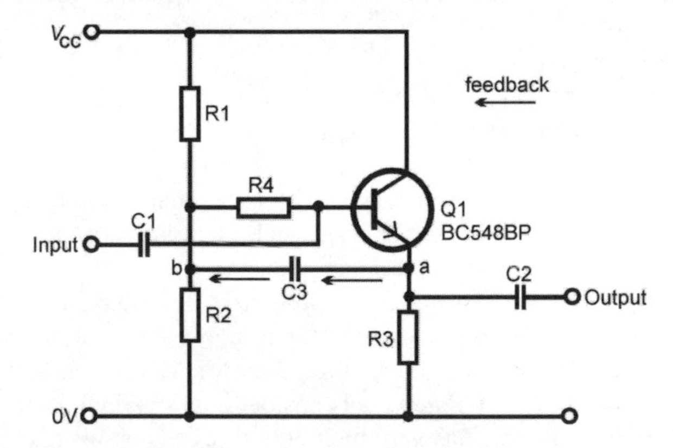

Figure 6.2 *Bootstrapping by the capacitor between nodes a and b gives this emitter-follower amplifier an input resistance of almost 500 kilohms, even though R1, R2 are as small as 15 and 18 kilohms, and R4 is only 3 kilohms.*

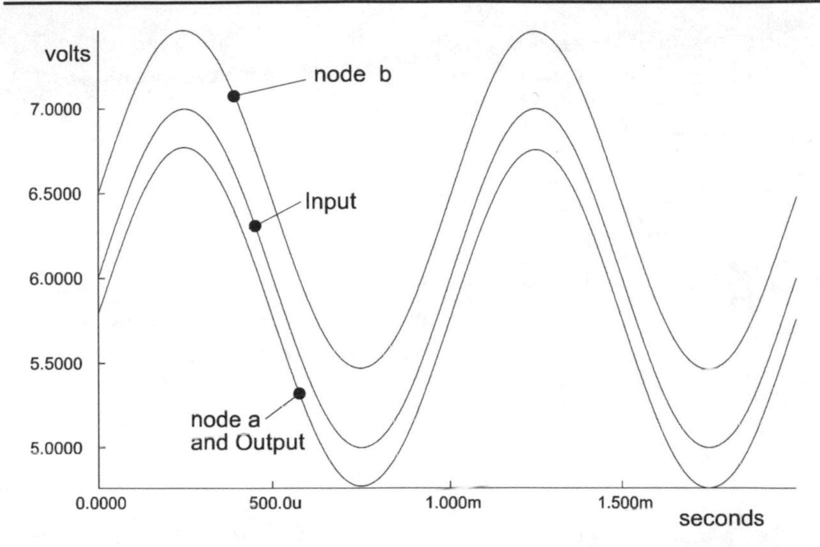

Figure 6.3 *These simulator plots of voltages in the amplifier of Fig. 6.2 illustrate the emitter-follower and bootstrapping actions.*

step is to add the capacitor C3 joining nodes *a* and *b*. Because of the action of the capacitor, any change of voltage at node *a* superimposes an equal change at node *b*. As Fig. 6.3 reveals, the voltage difference between *a* and *b* remains almost constant at about 0.7 V. To put it another way, the voltage at *b* is being forcibly altered by the output voltage – by its own bootstraps. Feedback does most of the work in pulling the voltage at *b* up and down and correspondingly less *current* is required from the input. Reducing the input current has the same effect as giving R4 a high value. Consequently, bootstrapping increases the effective input resistance. Measurements with a simulator show that even though the actual resistance of R4 may be only 3 kΩ, the input resistance of the bootstrapped circuit is now dramatically increased to almost 500 kΩ.

Darlington pair

This consists of two BJTs connected as in Fig. 6.4. Their collectors are joined to a common point which may be connected to the positive supply line. The emitter current of Q1 becomes the base current of Q2. As a result the current gain of the pair equals the product of the current gains of the two transistors. For example, if each transistor has a current gain of 100, the pair has a gain of 10 000. Only a very small base current is required by Q1, and this may be supplied by one or two biasing resistors of several megohms each. A common-collector amplifier based on a Darlington pair can have a very high input resistance.

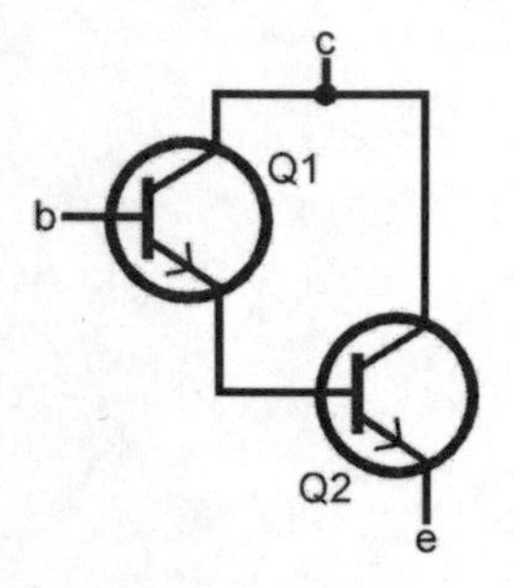

Figure 6.4 *A pair of transistors connected as pictured here is known as a Darlington pair. Because the emitter current of Q1 flows on to become the base current of Q2, the pair have a very high current gain.*

A Darlington pair may be constructed from two individual transistors, which need not be of the same type. For example, an emitter-follower circuit for driving a loudspeaker may have a low-power transistor for Q1 and a high-power transistor for Q2. Only the second transistor need be capable of providing enough current to drive the loudspeaker, with a consequent saving in cost. Another saving in cost is that only Q2 needs a heat sink. It is also possible for the two transistors to be enclosed together in a small package such as is normally used for single transistors. There are 3 terminal wires for collector, base and emitter, just as with a single transistor. Such a Darlington transistor can often be substituted for a single transistor, with consequent increase in current gain.

A similar device may be made from FETs and is known analogously as a *Fetlington* pair.

Common-base amplifier

A common-base amplifier feeds the signal into the emitter and takes the output from the collector (Fig. 6.5). The base is held at a steady voltage by a pair of biasing resistors (R_{B1} , R_{BI}). A by-pass capacitor C_{BP} shunts all a.c. signals to ground. The collector is held at a higher positive voltage by the pull-up resistor R_C and the emitter is held at low voltage by the pull-down resistor R_E. The collector is held close to the mid-rail voltage to allow the output signal to swing widely without distortion.

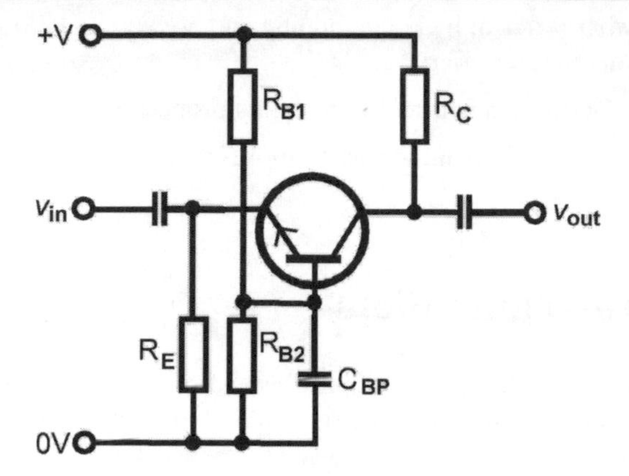

Figure 6.5 *A common-base amplifier has high voltage and power gain, unity current gain and low input resistance.*

The voltage gain of this amplifier is reasonably high, of the same order as that of a common-emitter amplifier, often somewhere between 100 and 150. But since the emitter current is only slightly more than the collector current, the current gain of this amplifier is very close to 1. Also, since the emitter current consists almost entirely of the collector current, it is a *non-inverting amplifier.* Unlike the common-emitter amplifier, it has a low input resistance, equal to r_e, which is only 25 Ω for an emitter current of 1 mA. Its output resistance is equal to R_C which is typically a few kilohms. The low input resistance limits the usefulness of the amplifier to circuits in which it is fed by a low-resistance source. For example, it may be fed from the secondary coil of a transformer, as it is in certain voltage-regulator circuits. We look at further examples later in this chapter.

The most advantageous feature of the common-base amplifier is that the base layer of the transistor isolates the collector region from the emitter region. As a result the capacitance between collector and emitter is very small and the amplifier works very well at high frequencies. We discuss high-frequency amplifiers in Chapter 8.

Keeping up?

1. What are the main features of the common-collector amplifier?
2. What is the main application of common-collector amplifiers?

3. Explain what is meant by bootstrapping and how it can be used in a common-collector amplifier.
4. What is a Darlington pair and what are its properties?
5. List the features of a common-base amplifier.

Differential amplifiers

A differential amplifier has two inputs instead of only one and its function is to amplify the *difference* between the two inputs. Usually we are concerned to amplify a small voltage difference between two relatively large voltages.

The most commonly used differential amplifier is illustrated in Fig. 6.6. Because the tail resistor (R_T) often has a high value, this kind of amplifier is also known as a *long-tailed pair*.

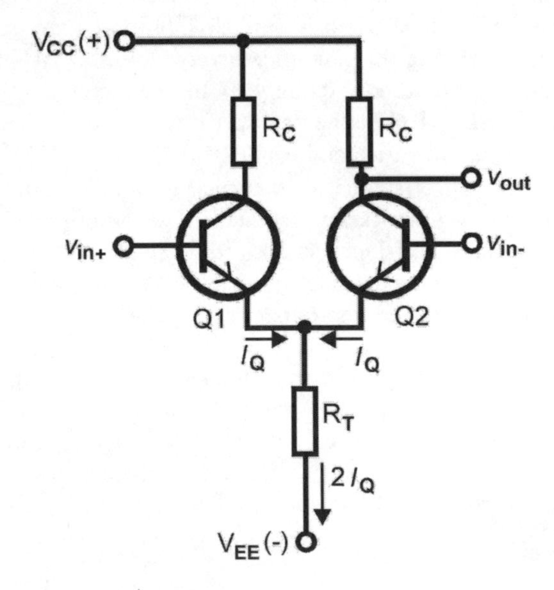

Figure 6.6 *A differential amplifier has two inputs and one or two outputs (one shown here). It operates on a dual supply. This circuit has several useful variants, as can be seen in the following pages.*

Before we describe the way this amplifier works, there are two terms that need explaining:

- *differential mode,* and
- *common mode.*

Differential mode describes the normal operation of the amplifier in detecting and amplifying the difference in potential between its two inputs. For example, the inputs may be connected to two electrodes in contact with the skin on a subject's forehead. A difference in potential detected at the electrodes indicates nervous activity in a certain part of the brain. But at the same time there are other potential changes caused by stray electromagnetic fields in the room. These arise from electrical equipment in the room, even electric lamps, telephones and other common items. The electromagnetic fields induce currents in the subject's body, including the head. These induced currents affect both electrodes equally. We say their effect is *common-mode.*

Another source of common-mode voltage changes can be friction between the subject's body, or parts of the body, against clothing, furniture and carpets. Under the right circumstances friction may raise or lower the body potential by several hundred volts. This raises or lowers the potential of both electrodes (and both amplifier inputs) equally by several hundred volts. Ideally, such changes should not affect the output of the amplifier. It should show *common mode rejection* and amplify only the difference between its inputs.

Common mode rejection implies that both sides of the amplifier are identical. They both have the same supply voltages and the resistors and transistors used in both sides of the amplifier should have identical performance. Often both transistors (and sometimes the collector resistors too) are all fabricated on the same chip. This eliminates any effects of manufacture and also differences in temperature between the two sides. If there are changes of ambient temperature both transistors are affected equally. The changes in their response are common-mode, are mostly rejected, and have a minimal effect on output.

Both sides of the amplifier are identical in structure but, because they are linked at the emitters, they act in opposite directions. Usually we take the output from the collector of Q2 but there is nothing to prevent us from taking output from the Q1 collector as well or instead. If we do this we find that the two output signals are equal in amplitude, as is to be expected in a symmetrical circuit. But the changes in output voltage are opposite in direction. When the output at Q2 rises the output at Q1 falls, and the other way about.

It helps us to remember what to expect if we decide that we will normally use only the output at Q2. Then we find that the output is always in phase with the v_{IN+} input. As the input at v_{IN+} rises and falls, the output at Q2 rises and falls too. The v_{IN+} terminal is described as a *non-inverting input.* A signal fed to this input is amplified without being inverted. The other input terminal, the v_{IN-} input, acts in the opposite direction. Its signal is amplified and inverted. It is the *inverting input*. We shall use these terms again in Chapter 10.

Common mode voltage gain

Fig. 6.6 shows that the amplifier requires a split supply with V_{EE} equal to $-V_{CC}$. If both inputs are connected to the same signal we find that the common-mode voltage gain is equal to $-R_C/R_T$. If we make both collector resistors (R_C) equal in value to R_T, the gain simplifies to –0.5. In words, the output varies by only half the variation in input. The negative sign shows that this is an inverting amplifier. Fig. 6.7 plots the input and output of such a differential amplifier, in which R_C equals R_T. The output amplitude is half that of the input and is inverted. We can vary the common-mode gain by varying the ratio between R_T and R_C. For example, if we make R_T equal to double R_C, the gain is reduced to –0.25. Although we can not reduce the gain to zero, we can certainly reduce it to a satisfactorily low level.

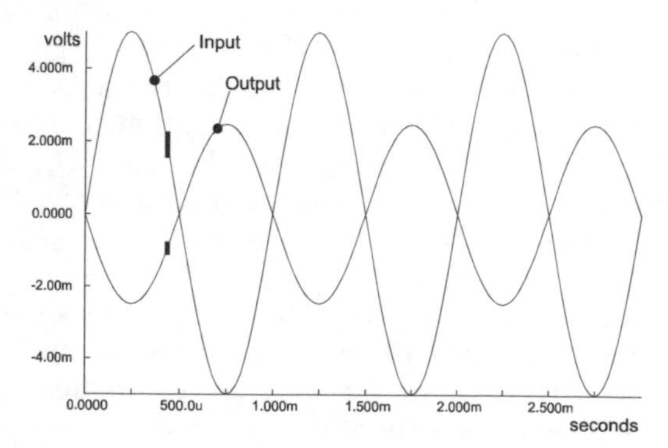

Figure 6.7 *A 1 kHz signal of 500 mV amplitude produces an output of 249 mV when fed to both inputs of a differential amplifier. The output is inverted, and demonstrates a common mode gain of -0.5.*

Differential mode voltage gain

Theory shows that if different signals are applied to each input, the differential voltage gain is approximately equal to $10i_TR_C$. Gain is positive if taken from the collector of C2, and negative if taken from the collector of Q1. Fig. 6.8 shows the output when two out-of-phase signals of the same amplitude are fed to the inputs. Note that these are plotted on a ×25 scale, as their amplitude is only 5 mV. The output amplitude is 906 mV, so the differential mode voltage gain is 906/5 = 181. Measurements show that i_T is constant at 192 μA. R_C is a 75 kΩ resistor. Applying the formula quoted above, the gain should be $10 \times 192 \times 10^{-6} \times 75000 = 144$, which is approximately equal to the gain measured on the simulator.

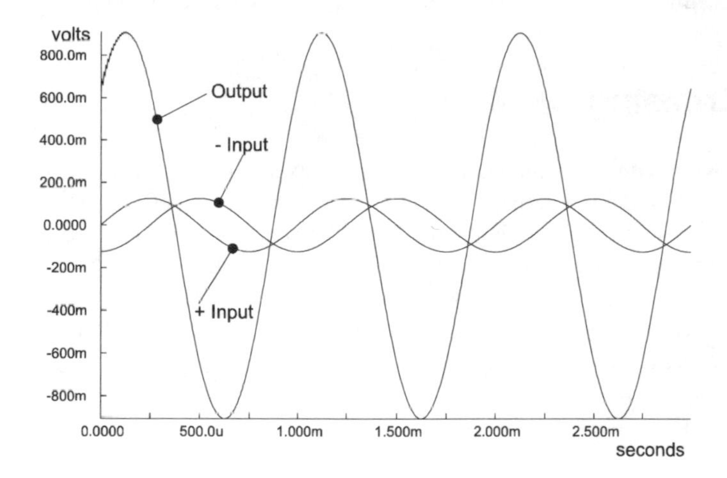

Figure 6.8 *At any instant the output is proportional to the difference between the inputs. Here, the inputs have amplitude 5 mV and are plotted on a 25 times scale.*

Common mode rejection ratio

A differential amplifier needs to have a differential voltage gain that is much larger than the common mode gain. We express the ratio between the two gains as the *common mode rejection ratio*, ignoring the signs of the ratios. In the example above, the common mode gain is 0.5 and the differential gain is 144. The ratio between them is:

$$\text{CMMR} = 144/0.5 = 288$$

In terms of symbols:

$$\text{CMMR} = 10i_T \times 2R_T/R_C = 20i_TR_T$$

This is assuming that R_T=R_C. The equation shows that we can obtain the desirable high value of CMMR by making either i_T or R_T (or both) as large as possible. But, according to Ohm's law, the current through R_T multiplied by the value of the resistor itself is equal to the voltage across R_T. This is a constant, approximately equal to V_{EE} (actually 14.4 V) and depends solely on the supply voltage used. Improving CMMR is therefore a matter of increasing V_{EE}, but this may be far from practicable or convenient.

Increasing CMMR

One way to do this which does not require excessively large values of V_{EE}, is to replace the tail resistor with a constant current sink (Fig. 6.9). The tail current now flows through Q3 and R_L to the negative rail. The base of Q3 is held at a constant voltage because V_{BB} is fixed and is stabilised by the Zener diode D1. Values are set to produce large collector current through Q3. This has the same effect as having a large negative value of V_{EE}, so increasing the CMMR.

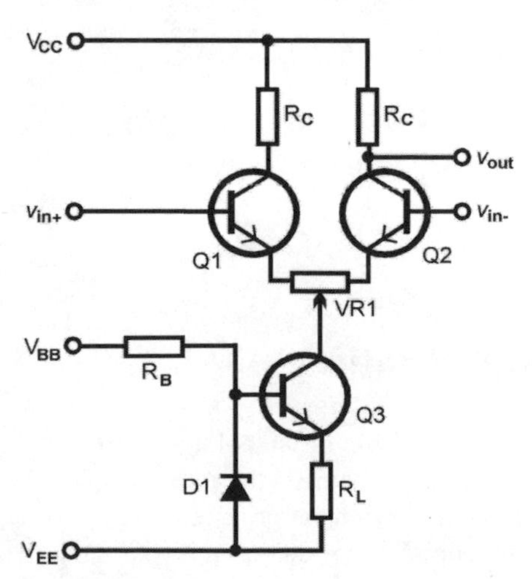

Figure 6.9 *Substituting a constant current sink for the tail resistor produces a higher common mode rejection ratio.*

A test on the circuit of Fig. 6.9 shows a common mode gain of only –0.001, while differential mode gain is just over 200. This gives a CMMR of 200/0.001 = 200 000.

Another refinement appearing in Fig. 6.9 is the variable resistor VR1. This allows the emitter currents of Q1 and Q2 to be balanced to compensate for slight differences in the gains of the transistors. The same principles apply if other transistor types replace the BJTs. Differential amplifiers with higher input resistance are built using MOSFETs or JFETs. We will look at other versions in Chapter 8.

Single-input differential amplifier

An interesting way of using the amplifier is to tie the inverting input to 0 V and apply the signal to the non-inverting input (Fig. 6.10). Since we have no intention of taking an output from Q1, the collector resistor of this transistor may be omitted. As might be expected, the signal is amplified without inversion.

Looking more closely at Fig. 6.10, we can see that the circuit is actually a two-stage amplifier consisting of an emitter-follower (Q1) feeding its input to a common-base amplifier (Q2). The re-drawn circuit diagram (Fig. 6.11) makes this clear. The emitter-follower provides high input resistance, which is a valuable feature for almost any amplifier. It is non-inverting, so

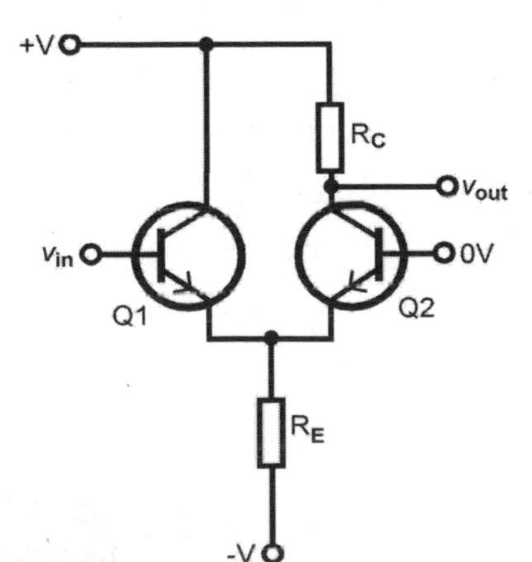

Figure 6.10 *For single-ended use the differential amplifier has its inverting input connected to 0 V. The collector resistor of Q1 may be omitted.*

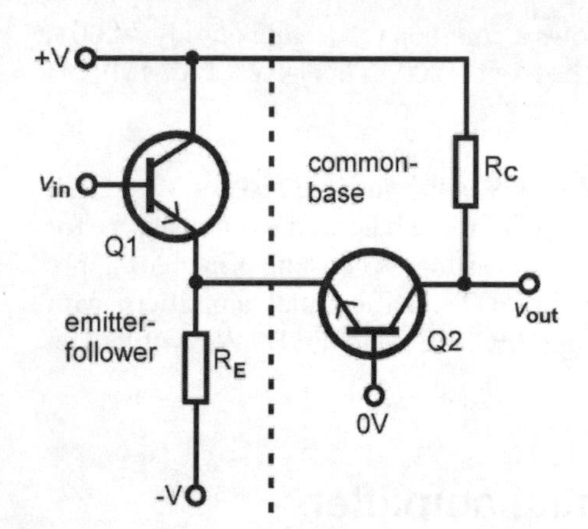

Figure 6.11 *Re-drawn, the differential amplifier of Fig. 6.10 can be seen to consist of two single-transistor amplifiers, each having advantageous properties.*

the signal it sends to Q2 is in phase with the input signal. The emitter-follower stage of the amplifier has unity voltage gain (approximately) but large current gain. In this way, the common-base amplifier receives the ample current, which is essential for its operation. The common-base amplifier shows no further current gain but has high voltage gain and fairly low output resistance. In combination, the two amplifiers have high input resistance, high voltage gain and reasonably low output resistance.

The whole amplifier is more than the sum of its two parts, for it also has the high CMMR that is characteristic of differential amplifiers. One input is the ground line, and the other might be connected to the output of a sensor, for example. We can minimise electromagnetic interference by using a cable with an earthed screen but sometimes there are practical difficulties in screening the whole length of the cable, particularly where the sensor leads are joined to the end of the cable. Interference will be picked up on both lines and will be largely eliminated as a common-mode signal.

This circuit, and also its variant with a constant current source, is a handy building block. If you inspect circuit diagrams of amplifiers, you will often find a differential amplifier as the input stage, though one of its inputs may be permanently wired to ground.

Using balanced outputs

Although we have previously assumed that the output levels of the differential amplifier will be measured relative to the 0 V rail, it is also possible to measure them with respect to each other. In other words, the amplifier output is the *difference* between the collector voltages of Q1 and Q2 (Fig 6.12). This has one immediately evident advantage. One output is the inverse of the other so, as one output goes up, the other goes down by an equal amount. The difference between the amplitude increases by double the amount. In other words, the amplifier has double the gain of one referenced to the 0 V line.

Since both sides of the amplifier are identical, equal input voltages should produce exactly equal output voltages. In theory, the CMMR should bc zero. The transistors and resistors are best fabricated close together on the same chip. Even if equality may be difficult to achieve in practice, we should expect an amplifier with balanced output to give superior performance.

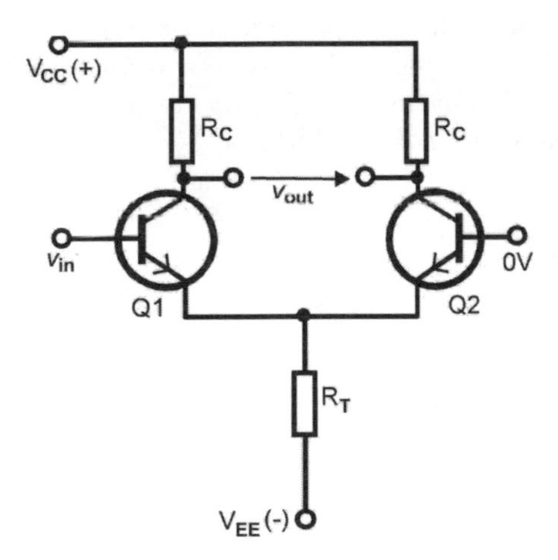

Figure 6.12 *The output of this differential amplifier is taken between the collectors of the two transistors. This arrangement gives double the gain of a normal differential amplifier.*

Linearity

Figs. 6.13 and 6.14 contrast the transfer characteristics of a single-transistor common-emitter amplifier with that of a basic differential amplifier. Both circuits use the same type of transistor, both are powered by a 15 V supply (±15 V in the case of the differential amplifier), and both havc 75 kΩ

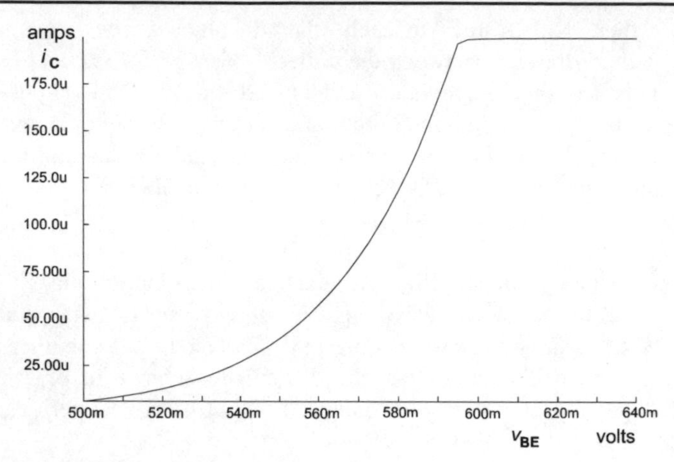

Figure 6.13 *The relationship between base voltage and collector current for a common-emitter amplifier. The response is far from linear. It flattens out at about 600 mV when the voltage drop across the collector resistor almost equals the supply voltage (15 V).*

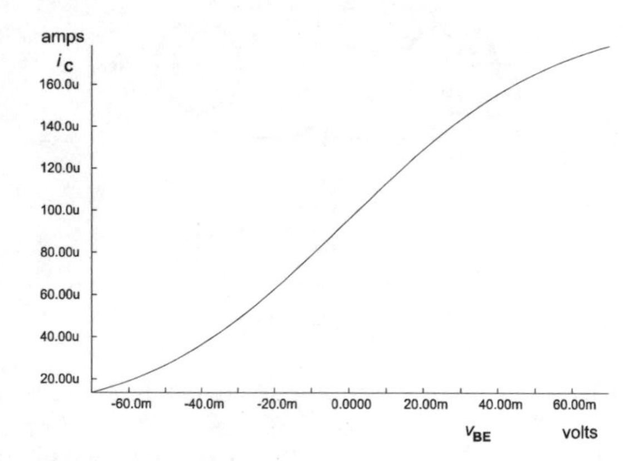

Figure 6.14 *In contrast to the response shown in Fig 6.13, that of a single-ended differential amplifier is much more linear.*

collector resistors. The graph in Fig. 6.13 plots the current in the collector resistor against v_{BE} as it is varied from 500 mV to 640 mV. This covers the same voltage range as a sinusoidal signal with offset 570 mV and amplitude 70 mV. As expected, current rises with increasing v_{BE} from about 13 μA to 178 μA, but the graph is by no means a straight line. The best we can do to compensate for this non-linearity is to keep signal levels small, say, less than ±10 mV. The graph also shows the transistor becoming saturated for signals greater than 600 mV. At this point the current through the resistor (about 180 μA) is sufficient to produce a voltage drop across it almost equal to the supply voltage and so the output is clipped. Clipping occurs at a relatively low v_{BE} in this circuit mainly because the value of R_C is high.

In Fig. 6.14 the input signal varies over the same range as in Fig. 6.13, and the output signal varies over almost the same range as before. But the response is much more linear and does not show clipping. Between - 40mV and +40mV the curve departs very little from a straight line. This means that signals of relatively large amplitude (up to ±40 mV) may be passed through this amplifier with relatively little distortion. The single-ended differential amplifier has superior performance to the basic common-emitter amplifier.

Output offset

The quiescent (no-signal) output of a differential amplifier is typically offset by a fraction of a volt with respect to the input. If we try to cascade several stages to obtain increased gain, the quiescent output is stepped up at each stage. This limits the number of stages we can conveniently cascade.

Increasing offset may be eliminated by using a capacitor to couple the amplifier to the next stage, but at the expense of introducing frequency-dependent effects. Another way of avoiding offset is to follow the amplifier with a second amplifier based on pnp transistors (Fig. 6.15). The offset in the npn amplifier is compensated for by an offset in the opposite direction in the pnp amplifier, so that the quiescent output voltage is zero.

We have seen that the gain of a differential amplifier is proportional to the value of R_C. In addition, the output resistance of the amplifier is *equal* to R_C. High gain implies a high value for R_C and therefore high output resistance. If we want the amplifier to have low output impedance we have to accept low gain. In this case we may cascade two or more low-gain amplifiers to give increased gain.

Using simple npn amplifiers introduces the problem of introducing output offset at each stage, so limiting the number of stages of gain. Cascading amplifiers like the one in Fig. 6.15 introduces no such problems.

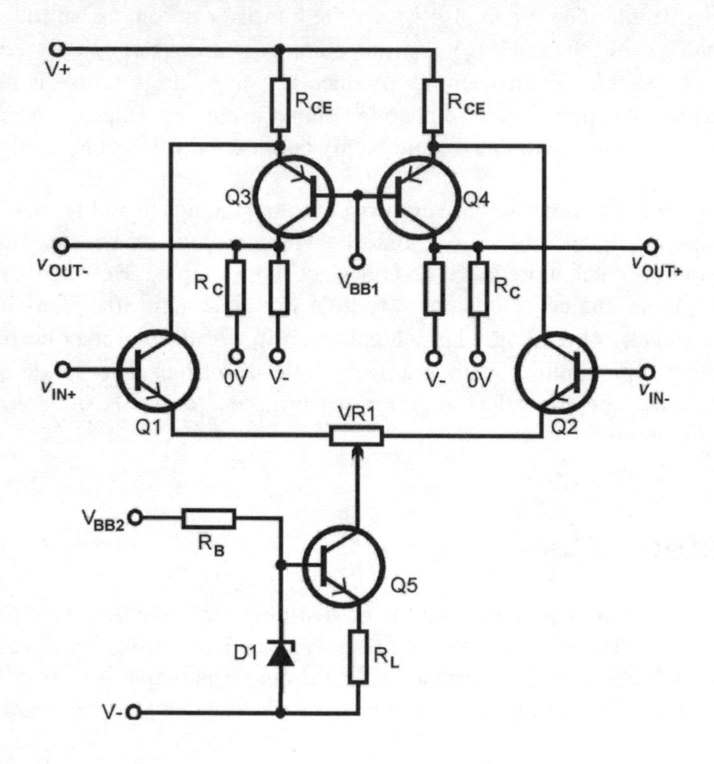

Figure 6.15 *A differential amplifier with a pnp output stage has zero quiescent output voltage.*

Keeping up?

6. Explain the difference between differential mode and common mode.

7. State typical common mode and differential mode voltage gains and define common mode rejection ratio.

8. What are the advantages of using a single-input differential amplifier instead of a common-emitter amplifier?

9. Explain what is meant by 'balanced outputs'.

10. Why is it necessary to cascade differential amplifiers if we require high gain?

Multi-stage amplifiers

We can obtain increased voltage gain by coupling two stages of voltage amplification as in Fig. 6.16. This shows two common-emitter amplifiers with the collector of Q1 wired directly to the base of Q2. The main concern is that both transistors should be working in their linear regions. This is done by choosing suitable values for the resistors. The collector and emitter resistors of Q2 are chosen to bring the output voltage (at the collector of Q2) as close as possible to the mid-rail value when the specified quiescent current is flowing. Then the collector and emitter resistors of Q1 are chosen to make the collector voltage of Q1 about 0.6 V greater than the base of Q2. Finally the biasing resistors of Q1 are picked to put the base of Q1 0.6 V higher than its emitter.

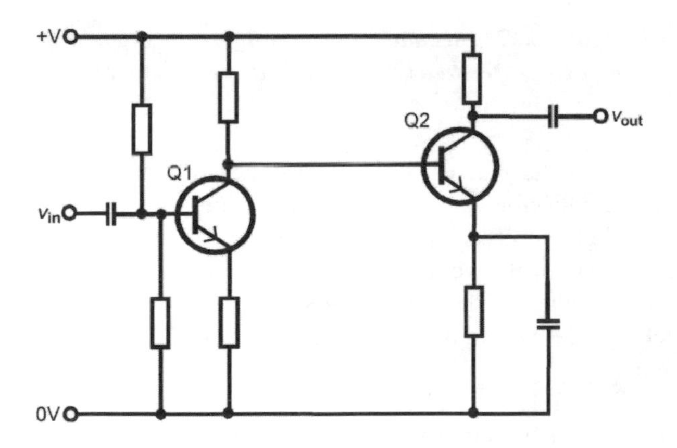

Figure 6.16 *An elementary two-stage amplifier.*

The two-stage amplifier provides several opportunities to improve its performance, that are not available in a single-transistor amplifier. As well as the within-stage negative feedback due to the emitter resistors, we can introduce inter-stage negative feedback. Fig. 6.17 shows the amplifier of Fig. 6.16 modified by supplying feedback along two routes.

Route A feeds back part of the output at the collector of Q2 to the emitter of Q1. The amount of feedback depends on the value of R1. If the characteristics of the particular transistor used for Q1, or the effects of temperature, or any other cause results in the collector current of Q1 being *higher* than it

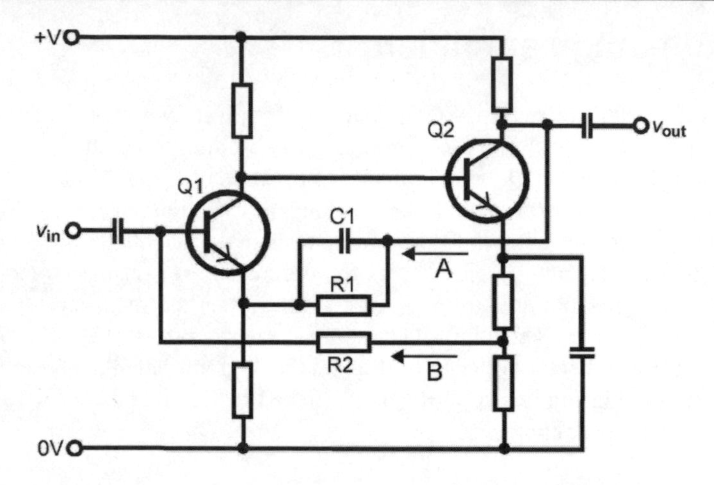

Figure 6.17 *This multistage amplifier uses two instances of negative feedback to improve its performance.*

should be, the voltage across the collector resistor increases. This results in a fall in the collector voltage of Q1 and in the base voltage of Q2. This turns Q2 slightly off, so raising the voltage at its collector. Feedback through R1 raises the voltage at the emitter of Q1, thus decreasing the base-emitter voltage. This results in a *decrease* of collector current. So feedback is negative and the operating conditions of the circuit are thereby stabilised.

Route B provides biasing for Q1, the original biasing resistors of Fig. 6.16 having been omitted. The biasing is now provided by tapping off part of the emitter voltage of Q2. This voltage is itself stabilised by the by-pass capacitor, performing the same function as C_{BP} in Fig. 2.16. A stable quiescent emitter current flows and the tapped voltage too is stable. A suitable value of R2 biases Q1 into its linear region. If for any reason the quiescent current of Q2 is too high, the tapped voltage is higher than usual, turning Q1 a little more on, and thus turning Q2 a little less on, so reducing the quiescent current. The reverse applies if the quiescent current is too low.

The input resistance of this amplifier is approximately equal to the value of R2. This is usually 100 kΩ or more, since only a small collector current is required in the first stage of the amplifier and therefore the biasing current can be small. The output resistance is the value of the collector resistor of Q2. This is usually fairly small, perhaps only a few kilohms, since the quiescent current through Q2 is usually large.

The frequency response is determined by the values of components in the feedback loops. The input capacitor and R2 make up a high-pass filter, the high value of R2 ensuring that the cut-off frequency is low, usually only a few tens of hertz. The high-frequency cut-off point at the output is determined by R1 and C1, which form a high-pass filter. At high frequencies, C1 acts as a by-pass capacitor. It routes the feedback signal directly to Q1, so reducing high-frequency gain. At low frequencies, less of the signal passes directly through C1, the total feedback is reduced and gain is normal.

The ease of stabilising is only one of the worthwhile features of a multi-stage amplifier. In Chapter 9 we see how the two-stage configuration helps to reduce noise.

Supply-line decoupling

Often the final stage in a multi-stage amplifier operates at high current. As the signal voltage fluctuates, it causes a relatively small but detectable replica of the signal to appear on the supply lines. Earlier stages are thus being powered by a fluctuating supply voltage and, in effect, the signal is taken and amplified again. Depending on the circuit and also on the stability of the power supply, feedback may be positive. The situation is rather like that of acoustic feedback, except that feedback is electrical, through the supply-lines. The amplifier becomes an oscillator!

The solution is to *decouple* the supply line from the output circuit. The most common way of decoupling the supply line is to wire a large-value capacitor (say, 100 μF) between the positive and ground rails. We also decouple the negative and ground rails if a dual supply is used. This holds the supply-line voltage steady, smoothing out any fluctuations due to large currents being drawn by the later stage. More than one decoupling capacitor may be needed if there are several stages. It is also common to wire a resistor between stages in the positive supply line itself. In conjunction with the decoupling capacitor, this acts as a low-pass filter and unwanted signals are passed through to ground.

The two-stage amplifier described above has direct coupling between its stages. If there are reasons why the quiescent collector voltage of Q1 can not conveniently be the same as the base voltage of Q2, we may couple the stages through a capacitor. This allows quiescent voltages to differ significantly, yet the signal passes freely from one stage to the next. In such an amplifier we need to bias the base of Q2 with one or preferably two additional resistors, in the usual way.

Inductors (interstage transformers) may also be used for coupling but the size and weight of inductors suitable for use at audio frequencies counts against them when portability is important, as is so often the case nowadays. Their use is mainly restricted to radio-frequency amplifiers.

Multi-stage amplifiers can also be built using FETs. Negative feedback is used in the same way as in BJT amplifiers to ensure stable operating conditions. Of greater interest are the improvements to be derived from bringing together an FET and a BJT in one multi-stage amplifier (Fig. 6.18). We can then take advantage of the beneficial features of each type. The advantage of the JFET common-source amplifier is its high input resistance. The pull-down biasing resistor, and therefore the input resistance of the circuit, can be 2 MΩ or more. But JFETs do not have high voltage gain, so we follow the JFET amplifier with a BJT common-emitter amplifier to provide the necessary voltage gain. In this example we use a pnp BJT. The collector resistor consists of two sections R1 and R2, of which R2 is much the smaller. A small portion of the output signal is tapped at the junction of R1 and R2 and is fed back to the source of Q1 by way of its source resistor and the by-pass capacitor C1. The function of C1 is to stabilise the gate-source voltage as described for the MOSFET amplifier of Fig. 2.16.

If the lower end of R1 is connected directly to the 0 V line, so that there is no feedback, the overall gain of the amplifier is around 50 times. But there is serious distortion of the waveform, which is clipped as output nears +V. Feedback is negative and serves to reduce fluctuations in the gate-source voltage. With the feedback loop completed, gain is stabilised and also reduced (to about 10) with a greatly improved shape to the waveform.

In a multi-stage amplifier the first stage is often planned as a low-noise stage, with small collector currents, or perhaps using an FET, which generates less noise than a BJT. The first stage passes a relatively noise-free signal to the high-gain second stage.

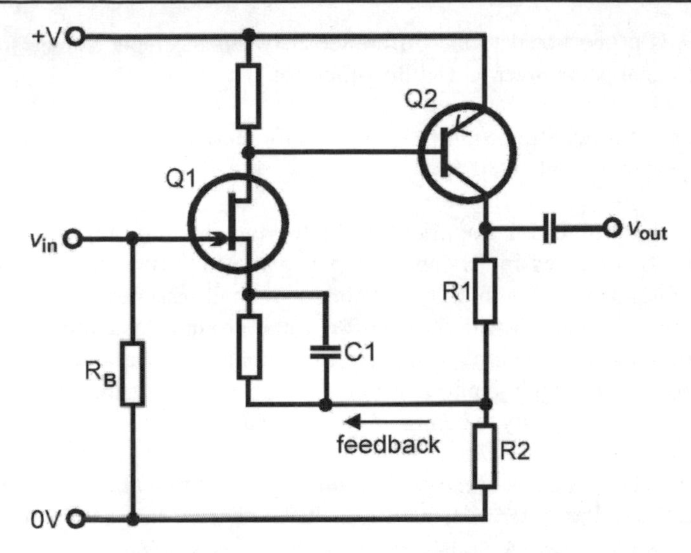

Figure 6.18 *The common-source amplifier has high input resistance and is followed by a pnp common-emitter amplifier to increase the voltage gain.*

Summing up

The common-collector (emitter follower) amplifier has a moderately high input resistance (but not as high as that of an FET amplifier) and low output resistance. It has a voltage gain of slightly less than 1 and is non-inverting, which makes it suitable for impedance (resistance) matching.

The effective input resistance of the common-collector amplifier may be increased by bootstrapping.

A Darlington pair has very high gain, so it can be used when an amplifier with high input resistance is required.

The common-base amplifier has low input resistance and fairly low output resistance. Its voltage gain is high but its current gain is only 1. It is a non-inverting amplifier. It has applications in circuits such as power supplies and also in high-frequency circuits.

A differential amplifier has two inputs and one or two outputs. Its output voltage is proportional to the difference between the input voltages, one output signal being inverted and the other not.

The common-mode rejection ratio is the differential voltage gain divided by the common-mode voltage gain.

Ideally, the differential amplifier has high common-mode rejection ratio. This can be obtained by having a large negative supply voltage or (more practicable) using a constant-current sink on the tail resistor.
A differential amplifier with its inverting input grounded is equivalent to a common-collector amplifier followed by a common-base amplifier. The combination has high input resistance, high voltage gain, low output resistance and immunity to common-mode signals.

If we take the voltage between the two outputs of a differential amplifier we obtain double the differential gain together with very low common-mode gain.

A differential amplifier has a much more linear response than a simple common-emitter amplifier. Its output signal is slightly offset. A differential amplifier with low output resistance necessarily has relatively low voltage gain, but several amplifiers may be cascaded to obtain higher gain. The use of amplifiers based on pnp transistors avoids offsets becoming too high.

Several stages of amplification may be coupled either directly, by capacitors or by inductors to make a multiple-stage amplifier.

Multiple-stage amplifiers provide increased voltage or current gain, but this is not their main advantage. The most important benefit is that we have the opportunity to use negative feedback from later to earlier stages in order to improve stability. Another advantage is that we may combine two or more types of amplifier to exploit the best features of each. In particular, we may use an FET amplifier with very high input resistance, followed by a high-gain BJT amplifier.

Test yourself

1. Describe what is meant by (a) thermal runaway, (b) bootstrapping, (c) small-signal current gain.

2. What is the advantage of basing an amplifier on a Darlington pair instead of a single transistor?

3. Which of the three basic types of BJT amplifier has (a) the highest input resistance? (b) the highest current gain? (c) a voltage gain close to 1? (d) the lowest input resistance? (e) the lowest output resistance?

4. In a Darlington pair a low-power transistor, $h_{fe} = 200$, is followed by a high-power transistor, $h_{fe} = 80$. What is the gain of the pair?

5. Describe the action of an npn differential amplifier. How do we obtain a high common-mode rejection ratio?

6. Describe the circuit of a two-stage amplifier using npn transistors. Explain all the ways in which negative feedback improves the performance of this amplifier.

7
Power amplifiers

A power amplifier is often used at the output stage of an audio system to drive the loudspeaker. Typically, power outputs of several tens or even a hundred or more watts are generated. But power amplifiers may also be used in other applications which need a high power output. Examples may be found in circuits for controlling electric motors, solenoids and high-intensity lamps.

A power amplifier may be simply one of the same basic amplifiers described in previous chapters, except that it is modified to handle the extra power. The small-signal transistor, such as the BC548, is replaced by one designed for passing larger current, such as the 2N3055. This is able to pass a current of 15 amps and operate at 115 watts. Lateral power MOSFETs are able to handle currents up to 16 amps and operate at up to 250 watts. Large currents inevitably mean that a considerable amount of heat is generated in the transistor. The current and power ratings are quoted on the assumption that the transistor will be mounted on a suitably efficient heat sink to dissipate the excess heat generated. In spite of the robustness of the transistor and the effectiveness of the heat sink, the transistor will still become very hot. Excess heat is one of the major limiting factors in power amplifiers. It is more of a problem with audio circuits and other circuits in which the transistor spends most of its time in an unsaturated state. In a switching circuit, such as one used in lighting control, the transistor is usually either fully off or fully on. When it is fully off, no current passes through it and no heat is generated. When it is fully on, there is a large current but the voltage drop across the transistor is a minimum, so the amount of power dissipated is small and heating is modest. Provided that the transistor is alternated rapidly from one state to the other, it can switch very high currents, perhaps several hundred amps, without becoming overheated. Special switching transistors with high power ratings are used on the output stage of switching circuits, but their use in such circuits relies on the switching being fast so that there is no time for the transistor to become overheated.

Avoiding high temperature is not just a matter of preventing the transistor from burning out. Temperature affects the performance of the transistor, so power amplifier circuits must be designed to be stable under a wide range of operating temperatures.

High power may be obtained by:

- generating high voltage signals,
- producing high currents, or
- a combination of these two techniques.

Generally the supply voltage of an amplifier is no more than about 20 V (excepting for valve amplifiers) and the power of the output is the result of the high output current. This suggests that the power amplifier should be one of the 'follower' amplifiers, either an emitter follower (common-collector) if we are using BJTs, or a source follower (common-drain) if we are using FETs. These have a low input resistance, combined with a very low output resistance. The required voltage amplitude is first generated by a low-power amplifier stage (or stages), often referred to as a pre-amplifier, and the output stage is a follower with high current output.

In this chapter we concentrate on BJT power amplifiers driving a loud-speaker. These illustrate the main principles of power amplification, which may be similarly applied to FET circuits and to other loads. The various types of power amplifier fall conveniently into five main classes, which are specified alphabetically. The types are defined according to the fraction of the signal cycle for which the transistor or transistors are switched on and conducting. This is determined by the amount of bias applied to the transistor. Exactly what this means and what effects it has will become clear as we discuss the various types.

Class A amplifiers

These are the simplest form of power amplifier and may consist of a straightforward common-collector amplifier (Fig. 7.1). A power transistor is used for outputs of more than few hundred milliwatts. For an amplifier of Class A, the transistor must be conducting for the whole of the cycle. It is biased sufficiently to keep it on continuously, even during the negative half of the cycle. For minimum distortion and maximum amplitude, the amplifier is run with a quiescent current sufficient to hold the output voltage

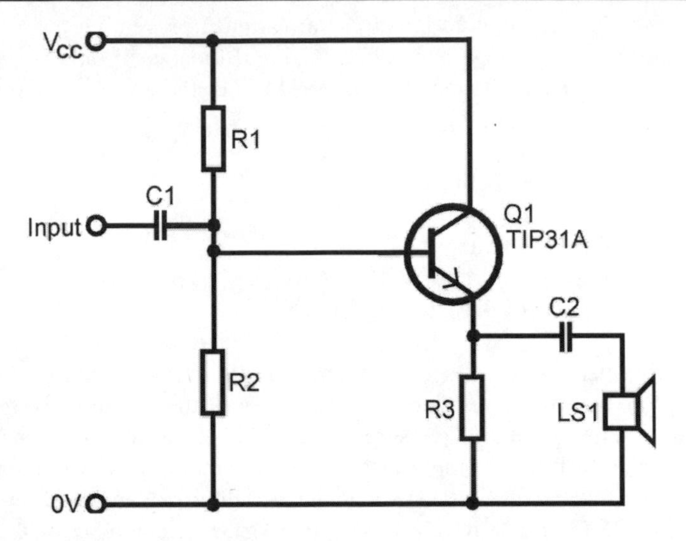

Figure 7.1 *This common-collector Class A power amplifier has a loudspeaker in parallel with the emitter resistor. The biasing resistors must have low values in order to provide sufficient base current.*

midway between the 0 V and positive rails. This allows the output to swing equally in either direction without distortion due to bottoming out or saturation. It follows that the amplifier is delivering 50% of its maximum power when there is no signal. This waste of power is one of the most serious disadvantages of Class A amplifiers. For example, analysis reveals that with typical component values the quiescent (no signal) current dissipation in R3 of Fig. 7.1 is about 2 watts, with a correspondingly large power dissipation in the transistor. Unless the transistor has an adequate heat sink there is a risk of thermal runaway.

Apart from this disadvantage, it can be seen from Fig. 7.2 that the output voltage follows the input voltage closely, as required for good reproduction of the sound. The maximum load current (that is, the current through the loudspeaker) is almost 500 mA, representing a power dissipation of about 1 W.

The output voltage follows the input voltage closely but, nevertheless, it does not follow it *exactly*. There is some distortion of the waveform. To estimate this we look at the results of a Fourier analysis (Fig. 7.3). In this figure the output voltage waveform of Fig. 7.2 is analysed mathematically

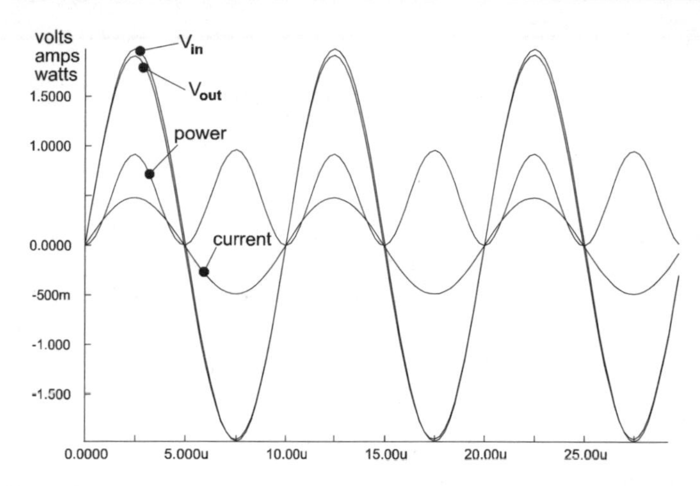

Figure 7.2 *The voltage output of the Class A amplifier of Fig. 7.1 closely follows the input. Note that the power in the load fluctuates at double the signal frequency.*

to determine its makeup in terms of pure sine waves of various frequencies. When analysed, the original input signal is seen to consist of the fundamental sine wave, which has a frequency 10 kHz in this example. Virtually no other frequencies are present.

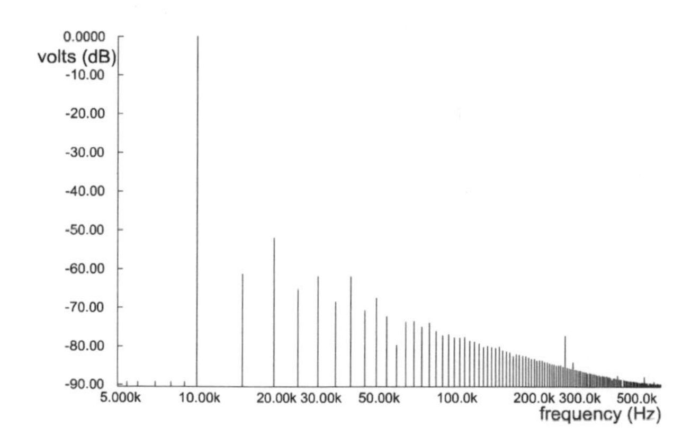

Figure 7.3 *A Fourier analysis of the output signal from the amplifier of Fig. 7.1 shows that it contains many harmonics, as well as numerous intermodulation frequencies (see box opposite).*

Fourier analysis

When a pure sine wave is amplified, it is most likely that the output will no longer be pure. The signal now consists of the fundamental (the original pure sine wave) to which have been added signals of other frequencies, usually with reduced amplitude. Most of these additional signals are related to the fundamental, having twice, three times, four times the fundamental frequency or even higher multiples. These are known as *harmonics.*

Harmonics occur naturally in certain sounds, such as those made by musical instruments, and are not then regarded as distortion. Indeed, the various harmonics and their relative amplitudes determine the characteristic timbre of each kind of instrument. But an amplifier can add harmonics of its own and these certainly give a distorted output.

The harmonics and other frequencies may also combine to produce *second-order distortion.* We find signals that have frequencies that are the *sum* of two of the frequencies already present, or that are the *difference* between two other frequencies. This is known as *inter-modulation distortion.* It may occur between the harmonics or when two or more unrelated frequencies are present in the original signal (as there usually are). Fortunately, the higher harmonics and inter-modulation frequencies usually have amplitudes appreciably less than that of the fundamental, so their effects are limited. Also the ear is insensitive to frequencies higher than about 20 kHz.

To study the various frequencies present in a signal we may use a *spectrum analyser*, an instrument which scans the signal to find which frequencies are present, and then measures their amplitude. The result are displayed on a tube resembling that of an oscilloscope. Given the graphical co-ordinates of the waveform, we can alternatively obtain a similar result by mathematical calculations. This is what happens when we run a Fourier analysis on a computer. Having used a simulator to calculate and plot the output waveform (as in Fig. 7.2), we next ask the computer for a Fourier analysis and obtain the plot (as shown in Fig. 7.3).

In the analysis of the output signal we see the 10 kHz fundamental represented by the tallest vertical line. As well as this we can see other lines representing other frequencies. For example there are lines at 20 kHz, 30 kHz, 40 kHz, and 50 kHz as well as even higher frequencies. These are the harmonics or 'overtones' generated within the circuit. Their presence indicates that the amplifier has distorted the signal to a certain extent. But the amplitudes of these signals are 50 dB or more down on the amplitude of the fundamental.

The decibel scale plots values as *fractions* of the 0 dB value, and –50 dB is equivalent to 0.00316 of the amplitude. So, if the fundamental has an amplitude of 4 V, the harmonic has an amplitude of 12.64 mV. In terms of power, the harmonic has a power of 0.00001 of the fundamental, for we square the *fractional* voltage to obtain the *fractional* power. There are many other frequencies present in Fig. 7.3, some of them being intermodulation frequencies (see box). Amplitudes mainly decrease with increasing frequency, and those below –60 dB (one thousandth) may be ignored, but there is one frequency (270 kHz) outstanding, possibly due to a resonance in the circuit.

If the strongest harmonic is rated at –50 dB with a power of only one hundred thousandth of the fundamental, we can accept this amount of distortion without unduly worrying about it. In general, Class A amplifiers show little distortion.

A number of variations on the Class A amplifier are possible. Instead of two biasing resistors we may have only one, either from the positive rail to the base, or from the collector to the base. These were described in our original discussion of the common-emitter amplifier. These arrangements give less stable operation than that which uses two resistors. Another simplification is to wire the speaker in place of R3, or between the positive rail and the collector. Increased current gain may be obtained by substituting a Darlington pair for the single transistor.

The class that an amplifier is assigned to depends only on the *conduction angle,* the portion of the 360° cycle of the signal during which the amplifying device is conducting. In Class A the conduction angle is 360°. The type of amplifying device used has no bearing on the class. Class A amplifiers may be based on BJTs, FETs, valves, Darlington pairs and several other devices.

Keeping up?

1. Which basic type of amplifier is used for power amplification?

2. What is the principle of operation of Class A amplifiers?

3. What precautions can be taken in the construction of a Class A amplifier to avoid the effects of excess heat production?

4. Explain the chief disadvantage of Class A amplifiers?

5. What is meant by the terms *fundamental* and *harmonic*?

6. If the power of signal A is only 1/100 000 of that of signal B, what is their relationship when expressed in decibels?

Class B amplifiers

The excessive waste of power of the Class A amplifier may be eliminated by using two transistors, one to handle positive-going excursions of the signal and the other to handle the negative-going excursions. This is known as *push-pull* operation. For each transistor the conduction angle is 180°. When there is no signal, both transistors are off and no power is being used.

Fig. 7.4 shows a basic Class B amplifier, which has an npn transistor for the positive-going phase and a complementary (matching) pnp transistor for the negative-going phase. Both transistors are configured as emitter-follower amplifiers. The result of applying a 4 V sine wave to the input is illustrated in Fig. 7.5. Output is zero when there is no signal because both transistors are off. But output stays zero for as long as input is less than about 0.6 V. This accounts for the distinct 'off' periods each time the input passes between +0.6 V and –0.6 V. These discontinuities in the output signal are known as *crossover distortion.* As might be expected, such marked distortion produces a degradation of sound quality which is very noticeable to the ear. A Fourier analysis of the output (Fig. 7.6) reveals that both even and odd harmonics are present, as well as a number of other frequencies. Note that, the amplitudes of the harmonics are much greater in the output of this Class B amplifier than they are in the Class A amplifier (Fig. 7.3). For example, the harmonic at 30 kHz is down only –20 dB (one tenth) on the level of the fundamental, compared with –62 dB (less than one thousandth) in the Class A amplifier.

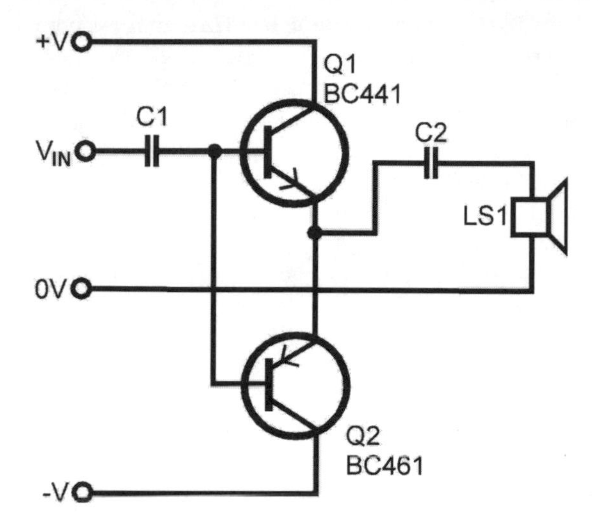

Figure 7.4 *A BJT Class B amplifier has two transistors, configured as emitter followers. Q1 (npn) amplifies current on the positive-going half cycles and Q2 (pnp) amplifies on the negative-going half-cycles.*

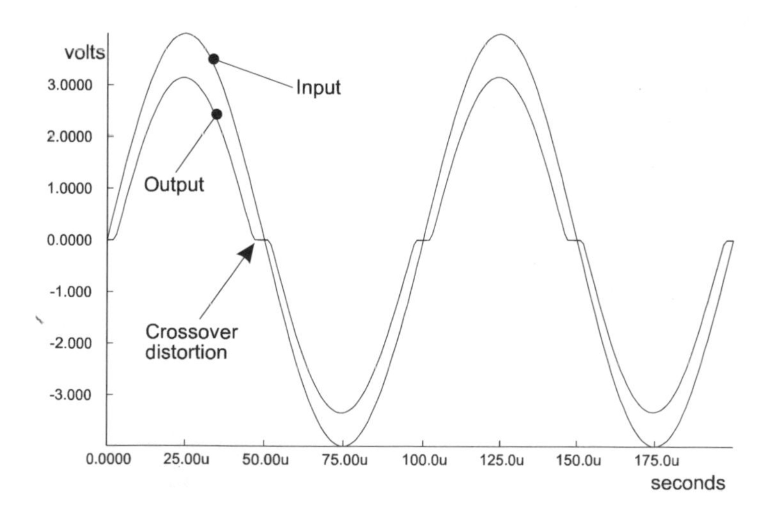

Figure 7.5 *The major fault of class B amplifiers is the crossover distortion which occurs as the input voltage swings through zero.*

The analysis of Fig. 7.6 is based on a 10 kHz signal for comparison with Fig. 7.3 but, if we are interested in the audio range, we can use a 100 Hz signal and plot the Fourier analysis from 30 Hz to 25 kHz (Fig. 7.7). Now we understand why the sounds from a simple Class B amplifier can be so unpleasant to listen to. Moreover, the reproduction is worse in the quiet passages. Fig. 7.7 is based on an input signal of 4 V amplitude. If we reduce the amplitude to 1 V (remembering that only input levels in excess of ±0.6 V produce any signal at all), the harmonics are relatively stronger. There are many stronger than -30 dB, which is 0.0316 of the amplitude of the fundamental signal.

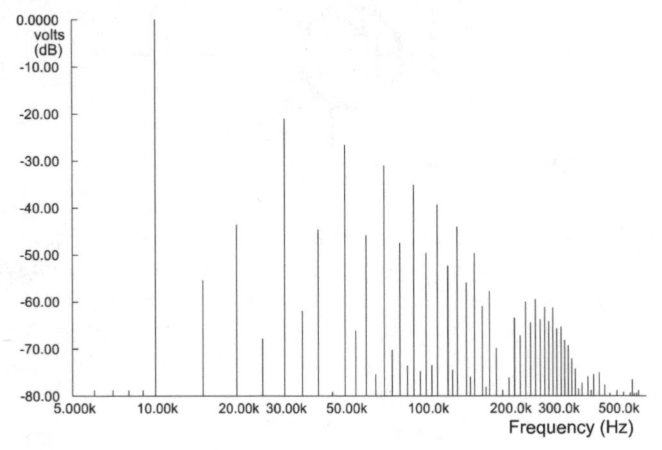

Figure 7.6 *A Fourier analysis of the output curve of Fig. 7.5.*

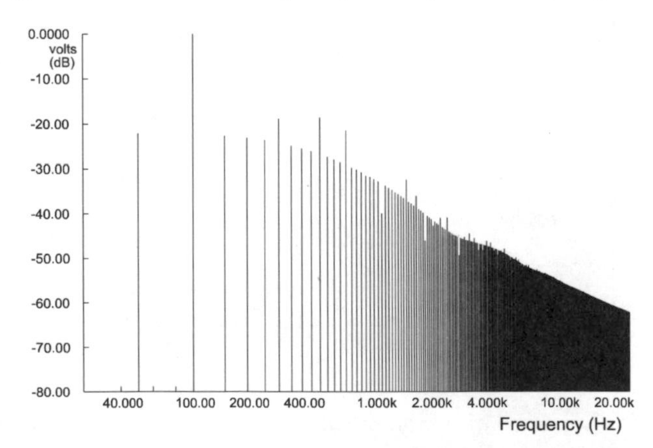

Figure 7.7 *Analysis of a 1 V, 100 Hz signal from the same amplifier.*

Diode drop and temperature

When a current is flowing through a pn junction, from p-type to n-type semiconductor (that is, the junction is forward-biased), there is a voltage drop across the junction. This voltage is about 0.6 V for a junction made in silicon, but only about 0.2 V in germanium devices. This voltage drop is a characteristic of a forward-biased diode, in which the p-type is the anode and the n-type is the cathode. The fall in voltage is often referred to as a 'diode drop'.

The same occurs in an npn transistor between the base region (p-type) and the emitter region (n-type). In normal operation, current flows from base to emitter so the base-emitter junction is forward-biased and there is a voltage drop of about 0.6 V in silicon BJTs. The exact value of the voltage drop depends on the temperature of the junction. For example, in the amplifier of Fig. 7.1, the base-emitter voltage is 0.78 V at 0 °C but only 0.53 at 150 °C (a very hot transistor). Variations such as this alter the quiescent current obtained by the biasing and hence the quiescent collector current of the amplifier. A test run on the Class B amplifier showed the power output increasing from 1.18 W at 0 °C to 1.36 W at 150 °C.

Temperature stabilising

One way to compensate for the temperature-dependent variation in the base-emitter voltage is to introduce a pair of diodes into the amplifier (Fig. 7.8). The circuit also includes resistors R3 and R4 to reduce the consequences of differences between transistors. Current flows through the chain R1-D1-D2-R2, which is symmetrical and the no-signal voltage at the point between the diodes is 0 V. The diode drop across the diodes brings the base of Q1 to about 0.6 V, and that of Q2 to about –0.6 V. We are solving two problems at once here, for the fact that the bases are at these voltages means that the transistors are *on the verge* of being turned on. Immediately the input signal rises above 0 V, the base of Q1 rises above +0.6 V, Q1 is turned on, and begins to conduct. Similarly, the instant the input signal falls below 0 V, the base of Q2 falls below –0.6 V, Q2 is turned on, and begins

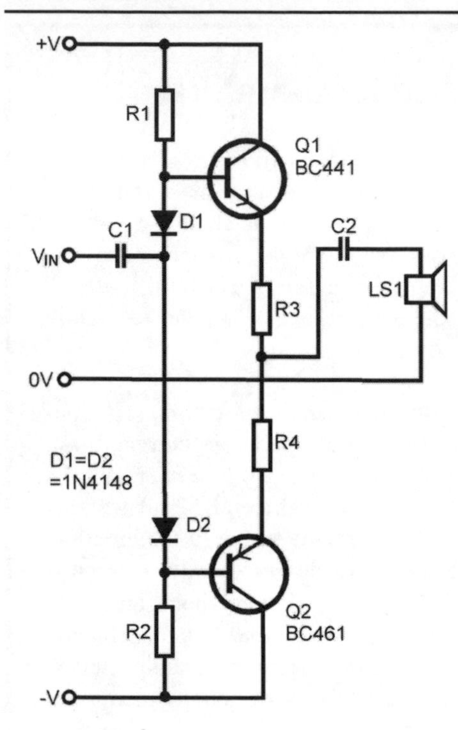

Figure 7.8 *The diodes in this improved version of the Class B amplifier serve to bring the transistors to the point of being just switched on when no signal is present. The fact that they have a pn junction also compensates for variations in the temperature of the transistors.*

to conduct. This arrangement eliminates the discontinuity at crossover, so crossover distortion does not occur. The circuit is therefore a compromise between Class A (in which the transistor is fully on at all times) and Class B, (in which the transistors come on only when the signal exceeds ±0.6 V). This arrangement is often referred to as Class AB.

Examination of the output from this circuit on a simulator shows an apparently pure sine wave, with no discontinuities at crossover. A Fourier analysis of the output of a 100 Hz signal confirms this. Apart from the 100 Hz fundamental, the first distortion frequency present is 300 Hz, with an amplitude of –100 dB.

To return to the problem of temperature stabilising, the variation of base-emitter voltage with temperature is compensated for by the variation of diode drop with temperature. In both the diodes and the transistors there is a forward-biased pn junction, so changes in base-emitter voltage are matched by changes in diode drop, particularly if the transistors and diodes are mounted on the same heat-sink. If temperature increases, reducing base-emitter voltage and thus tending to increase base current flow to the transistor, the diode drop too is reduced and pulls the base voltage down. Analyses of the output at temperatures ranging from 0°C to 150°C show no perceptible temperature effect.

Dual npn amplifier

A Class B push-pull amplifier is usually based on a complementary pair of transistors. If the output is to be symmetrical, the transistors must have equal gain. Matching pairs of npn and pnp transistors for use in audio amplifiers are sold expressly for this purpose. On the other hand, it is easier to manufacture a pair of identical transistors if they are *both* npn or *both* pnp. The circuit of Fig. 7.9 uses a pair of npn transistors. They are connected in parallel across the supply lines and both fed by the same biasing network. This is wired to the centre-tap of the secondary coil of the input transformer. This coil receives the signal inductively from the primary coil. At any instant the two ends of the secondary are of opposite polarity. When the 'upper' end (in Fig. 7.9) is positive, the upper transistor is conducting. At the same time the 'lower' end is negative and the lower transistor is turned off. In this way we obtain the same kind of action as in

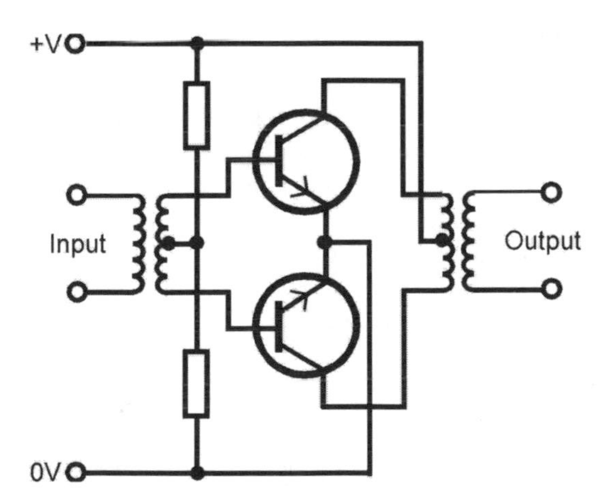

Figure 7.9 *Instead of a complementary pair, this amplifier uses two npn transistors. A centre-tapped transformer provides base current to each transistor alternately and a similar transformer is used to combine their outputs.*

Fig. 7.4. The output signals from the collectors are fed to the primary of the output transformer. The centre tap of this is connected to the positive supply so, at any instant, either one or the other half of the coil (but not both halves) are carrying the signal. The secondary coil picks up the signal inductively from either half of the primary and the output appears between the secondary terminals.

Keeping up?

7. What is meant by *push-pull* operation?
8. What type of BJT amplifier is used in a Class B amplifier?
9. Why is a Class B amplifier economical of power?
10. What is the chief disadvantage of a Class B amplifier?
11. How do the diodes in the circuit of a Class AB amplifier improve its performance?
12. Describe a Class B amplifier which does not use a complementary pair of transistors.

Class C amplifiers

Continuing the progression of smaller and smaller conduction angle from A to AB and then to B, we come to Class C in which the conduction angle is appreciably less than 180°. This mode of operation is obtained by biasing the transistor well below cut-off. It does not begin to conduct until the signal exceeds a certain level, so that only the peaks of signal are amplified. Because the transistor is conducting for only a fraction of the time, this type of amplifier is highly efficient (over 80%).

Class C amplifiers are used only at radio-frequencies. Fig. 7.10 illustrates the main features of a simple amplifier of this class. Essentially, this is a common-emitter amplifier similar to that in Fig. 5.7, except that there are no biasing resistors and there is an inductor L1 in place of the collector resistor. Some amplifiers in this class may be biased, but other are not. In this example the signal is a sinusoid with 0 V offset, and has an amplitude sufficiently great to take the transistor into conduction on its positive peaks. At low frequency the inductor has only low impedance. It acts as a

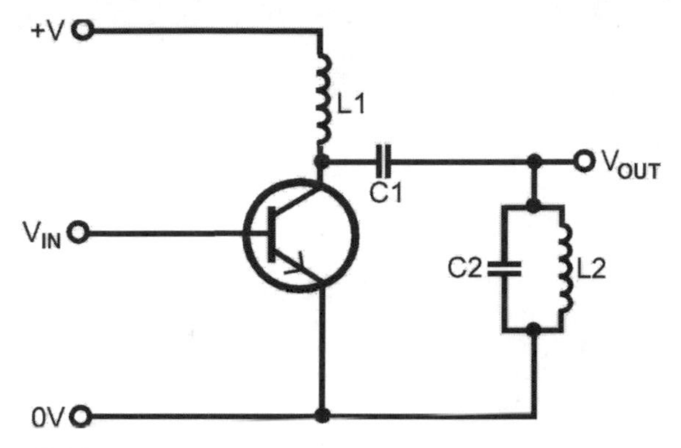

Figure 7.10 *In a Class C amplifier the output comes from the tuned network, C2/L2. It resonates at the same frequency as the input signal, which is amplified for only a part of its cycle.*

low-value resistor and the signal levels developed across it are small. At high frequencies the impedance of L1 rises to several kilohms and the signal across it is much larger. The value of L1 is chosen to bring the quiescent signal level about half-way between the supply rails when the signal is at the intended operating frequency. This frequency would normally be that of a carrier wave which is then modulated in amplitude or frequency by an audio or other signal. Using L1 in this way is typical of radio-frequency amplifiers, and not a particular feature of Class C.

Because the conduction angle is less than 90° and only the peaks of the signal appear at the collector of the transistor, the output signal is badly distorted. For this reason Class C amplifiers are totally unsuitable as audio amplifiers. Fig. 7.11 shows signals typical of the amplifier of Fig. 7.10. The upper curve is the input signal at 1 MHz, amplitude modulated at 100 kHz. Normally the modulating frequency would be far less than this but we have used this value in the simulated circuit from which Fig. 7.11 was produced so that we can show the RF signal *and* its modulation on the same time-scale. Note that the upper curve is plotted on a voltage scale that is ten times that of the other two curves. The middle curve shows the voltage at the collector of the transistor. A considerable amount of amplification has occurred. However, the signal consists of a series of peaks, rising above the 0 V level, and there are no negative-going peaks. The frequency is the same as that of the input signal and the amplitude of the peaks is proportional to that of the input signal.

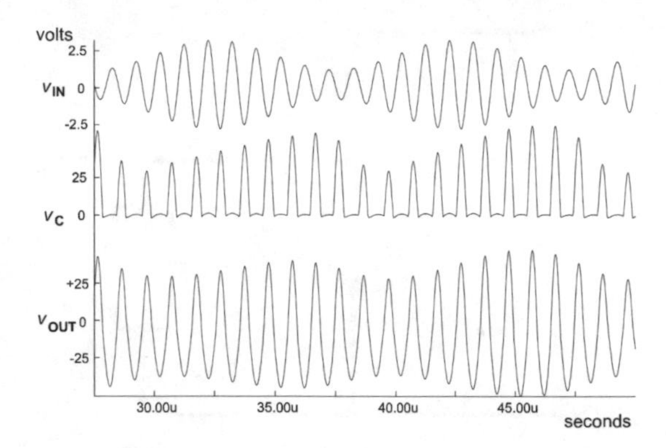

Figure 7.11 *An amplitude-modulated sinusoidal signal (v_{IN}) is amplified by a Class C amplifier to produce a very distorted waveform (v_C) at the collector. The tuned network restores the sinusoidal shape to the output signal v_{OUT}.*

A special feature of the Class C amplifier is the tuned network R2/C2 at the output. The values of R2 and C2 are chosen so that the network resonates at the carrier frequency. The way the circuit works may be likened to pushing a child on a swing. The swing has its natural frequency of oscillation, dependent upon its length. The motion of the swing, as measured by the horizontal displacement of the seat from its centre of travel, gives a sinusoid when plotted against time. To keep the child swinging, we push briefly each time the swing is moving away from us. If we apply a short push at the same time in each cycle (that is, at the natural frequency of the swing), we can keep the swing in motion indefinitely. This is just what happens in the amplifier. The LC network is kept in oscillation by regularly-repeating pulses passing through C2 from the collector of the transistor. If the pulses become larger the network oscillates with greater amplitude. This is just the same as when we push *more strongly* on the swing (but still for the *same short time* and at the *same frequency*) and make the child swing higher. Although our pushes are intermittent and relatively jerky, the motion of the swing is smoothly sinusoidal. In the amplifier, the LC network receives a pulsed input and produces a sinusoidal output. This can be seen in the lower curve of Fig. 7.11. The curve is not a perfect sinusoid but it is a great improvement on the signal at the collector.

Designing Class C amplifiers is a tricky matter in practice. Although the basic idea is simple enough, many complications arise when the circuit is put together on the circuit board (or on the simulator). At high frequencies the presence of stray capacitances and inductances produces effects that are often difficult to anticipate. Component values need careful trimming to get everything just right. Also there are so many possible interactions between these capacitances and inductances that it is almost impossible to reset one value without having to readjust others.

We have demonstrated the circuit amplifying an amplitude-modulated signal, but it can also be used to with frequency-modulation. Unfortunately we can not illustrate this by a set of plots like Fig. 7.11 because it requires appreciable percentage frequency modulation to produce a waveform like Fig. 1.3 that clearly shows the variation in frequency. Varying the frequency to such an extent takes the carrier signal outside the tuned range of the amplifier. In practical radio transmission, with a carrier frequency of, say, 1 MHz, the modulating audio signal has a maximum frequency of about 20 kHz. The signal frequency thus varies between 1.02 MHz and 0.98 MHz, both extremes lying well within the pass-band of the amplifier. As long as the signal does not vary too widely from the tuned frequency, it can force the LC network to oscillate at nearby frequencies and so follow variations in the frequency of the input.

An alternative and rather simpler arrangement is to omit the inductor L1 and to locate the RC network in its place. Output is taken from the collector of the transistor.

The amplifier described above belongs to the *non-saturated* type of Class C amplifier. While the transistor is conducting it is never fully saturated. The second type in Class C is the *saturated* amplifier. The biasing is set so that the transistor becomes fully saturated on the signal peaks, producing an almost square wave at the collector. Now the LC network receives pulses that vary in *length* instead of *amplitude*. The effect on the network is the same. As pulse length increases (frequency remaining constant, or in the case of FM varying only slightly), the network oscillates with greater amplitude. In other words, it turns a square-wave signal of varying mark-space ratio into a sinusoid of varying amplitude. Referring to the analogy of the child on the swing, instead of pushing with varying amounts of force, we push with *equal* force at every cycle but push for varying lengths of time. In terms of mechanics, we are giving energy to the swing by a succession of impulses. The impulse of a force is defined as (force applied) × (time for which force is applied). In the unsaturated amplifier we vary the force and in the saturated amplifier we vary the length of time.

The saturated amplifier is even more efficient because, for as long as the transistor is driven into saturation, there is no further increase in the power dissipated by the transistor.

Class D amplifiers

These amplifiers operate in switch mode, making them totally unlike the amplifiers of Classes A, B and C. They use transistors as switches, turning them fully on or fully off, with a rapid transition between one state and the other is very efficient. As we mentioned at the beginning of this chapter, a transistor may be operated at much higher power in switch mode. A typical Class D amplifier based on BJTs is illustrated in Fig. 7.12. The transistors are connected in parallel as common-emitter amplifiers, with each with half of the winding of T2 in their collector circuit. Current to their bases comes from the half the secondary winding of T1. The signal source is connected to the primary of T1 and the output is taken from the secondary coil of T2. It is arranged that the transistors are driven into saturation alternately, producing a square-wave signal into the load. The mark-space ratio of the square wave depends on the amplitude of the input signal. The input signal can have any shape provided that there is a fast transition through the switching voltage. In Fig. 7.13 we see the input signal, a modulated 1 MHz sinusoid, and the square-wave output from the transformer T2 when the

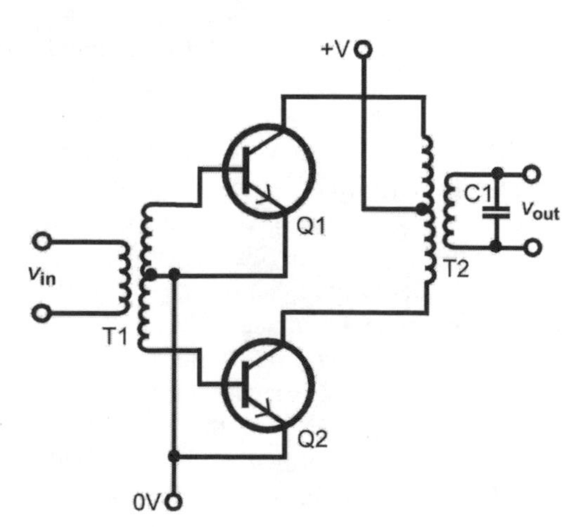

Figure 7.12 *In a Class D amplifier the transistors are switched fully off or fully on. The square-wave output from T2 may be converted to a sinusoid by C1.*

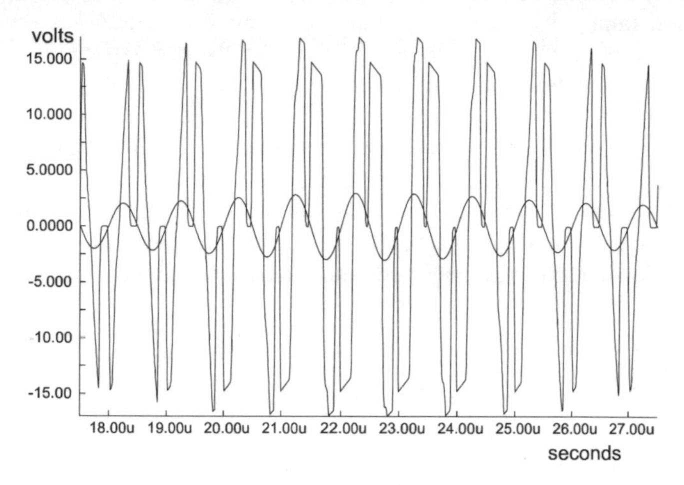

Figure 7.13 *Without the capacitor, the output of Fig. 7.12 is a square wave of varying mark-space ratio. In this circuit, it includes a strong 2f harmonic.*

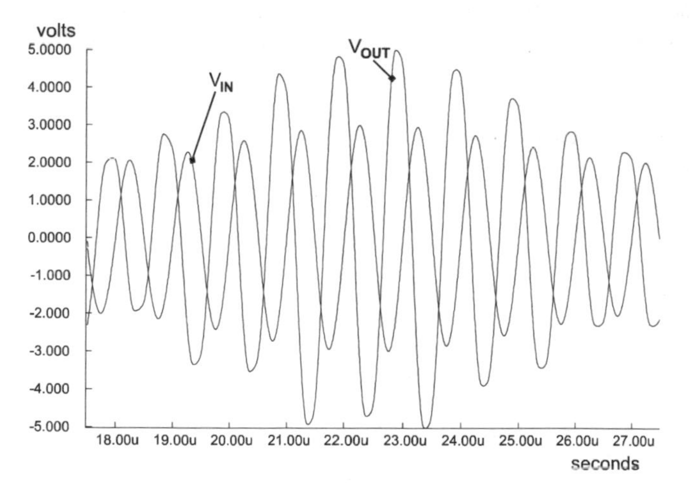

Figure 7.14 *The addition of the capacitor shown in Fig. 7.12 restores the output signal to an amplitude-modulated sinusoid.*

capacitor C1 is omitted. Note that a harmonic signal has been introduced, probably because of capacitances within the transistors (this is a simulation, but in a real circuit there would no doubt be other harmonics, due to stray capacitance and inductance elsewhere). The square-wave output has a constant amplitude because the transistors are driven into saturation at every cycle. But it can be seen in the figure that the mark-space ratio varies with the *amplitude* of the input signal.

Including the capacitor in the circuit forms an LC network with the secondary coil of T2. This has a sinusoidal output, as shown in Fig. 7.14. Whether or not the original signal was sinusoidal, triangular or some other shape, we now have a reasonably pure sinusoidal output at the same frequency, and this has amplitude-modulation which matches the input signal.

Fig. 7.15 is another Class D amplifier, this one being based on a pair of MOSFETs. This is a half-wave amplifier, since output is taken from only one of the MOSFETs. Both amplifiers have efficiencies approaching 100% since the transistors are acting as switches. The only losses are in the resistance of the transistors when ON and the capacitance within the transistors. An additional feature is that, by setting the resonance of the output network, it may be made to oscillate at one of the harmonic frequencies. For example, Fig. 7.13 showed that the $2f$ harmonic is present and the amplifier could be made to produce an output at this frequency, so acting as a *frequency multiplier*. The energy of the fundamental is not lost, for it drives the circuit at $2f$ instead. The swing analogy still holds good, for a swing may be kept going with a push at each alternate cycle (or every third cycle, or fourth cycle, and so on) instead of at every cycle.

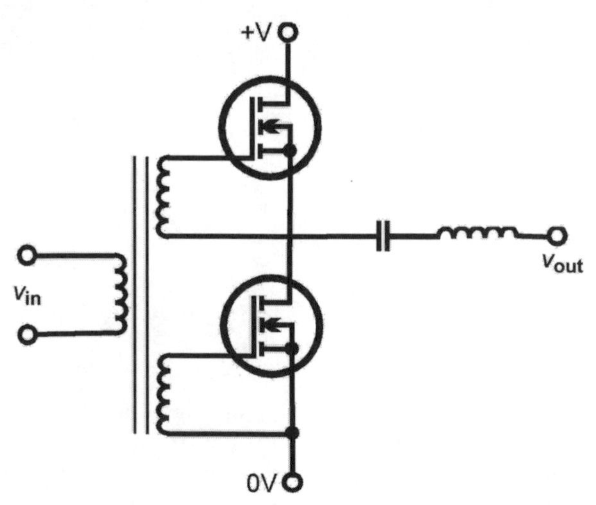

Figure 7.15 *A Class D amplifier based on a pair of MOSFETs.*

Summing up

Power amplifiers make use of transistors rated to operate at higher currents, and they are mounted on heat sinks to dissipate excess heat.

Typical power amplifier circuits are based on the emitter-follower or source follower configuration.

Power amplifiers are classed according to the conduction angle, the fraction of the cycle of a signal for which each transistor is switched on.

Class A: A simple follower amplifier. Conduction angle = 360°. Low distortion. Low efficiency.

Class B: Uses two follower amplifiers, usually complementary, one for each half of the signal phase. Conduction angle = 180°. Subject to crossover distortion. Diodes are used to stabilise the amplifier against temperature changes and at the same time eliminate crossover distortion (Class AB).

Class C: Conduction angle < 90°. High distortion. Used at radio-frequency. Includes a tuned LC network that is kept in oscillation by the peaks of the signal and re-creates a sinusoidal output at the carrier frequency. Can be used for amplitude-modulated or frequency-modulated signals.

Class D: Almost 100% efficient. Operates in switch mode and produces a sinusoidal output. Can be used as a frequency multiplier.

Test yourself

1. For each of classes A, B, AB, C, and D sketch the circuit of a typical amplifier, describe how it works, and sketch the output signal produced by an amplitude-modulated input signal. What is the distinctive feature of each class? Suggest one or more applications for each type of amplifier.

2 An input signal consisting of two pure sinusoids, 1 kHz and 2.5 kHz, is passed into an audio amplifier. What frequencies might we expect to find in the output signal?

8
High-frequency amplifiers

If we apply a signal of constant amplitude to an amplifier and vary its frequency, we find that the output amplitude falls off at high frequencies. For example, taking a common-emitter amplifier of the type illustrated in Fig. 5.7, we can plot a graph of the amplitude as frequency increases. For a given set of component values and with a sine wave input of 10 mV amplitude, the output varies as plotted in Fig. 8.1. Output amplitude remains constant at 1.7 V for all frequencies from 100 kHz up to about 3 MHz.

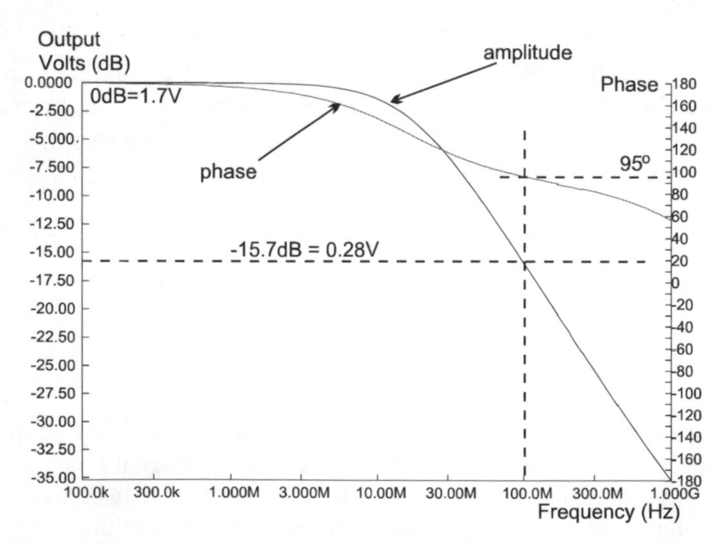

Figure 8.1 *The output amplitude of a common-emitter amplifier falls steadily when frequency exceeds about 3 MHz. At 100 MHz the output amplitude has fallen from 1.7 V to 0.28 V, and its phase lead is 95º, instead of 180º.*

Decibel scale

This is a measure of the ratio between two quantities. For two quantities x_1 and x_2, the ratio between them in decibels is represented by n, where:

$$n = 20 \times \log_{10}(x_2/x_1)$$

Note that the two terms in x must both have the same units. In the example of the amplifier which produced Fig. 8.1, we compare amplitudes with the maximum amplitude, that is 1.7 V, so $x_1 = 1.7$. Then, supposing we measure the amplitude at another frequency, (say 100 MHz) and find it to be 0.28 V, then $x_2 = 0.28$. From these two values we calculate $n = 20 \times \log_{10} (0.28/1.7) = 20 \times \log_{10} 0.165 = -15.7$ dB. The amplitude at 100 MHz is –15.7 dB in relation to the amplitude at low frequencies.

The plot for amplitude on a decibel scale against frequency on a logarithmic scale is known as a *Bode plot* and is widely used as a way of illustrating the frequency response of an amplifier or other frequency-dependent electronic circuit. In the example given, the amplifier has a voltage gain of 170 at frequencies up to about 3 MHz.

Above 3 MHz the amplitude decreases, slowly at first and then more rapidly. To allow a wide range of amplitudes to be plotted on the same graph we have plotted amplitude on a decibel scale. For the same reason, we have plotted frequency on a logarithmic scale. We have also plotted phase, which we shall discuss later.

Amplifiers can be classified according to the nature of their frequency response. A *DC amplifier* operates with full gain, provided that the input voltage is constant (DC) or is changing only slowly. DC amplifiers are often designed for high precision and stability and it does not matter that the gain falls off at a relatively low frequency. The gain of a typical DC amplifier reaches the –3 dB level when frequency is about 50 Hz, and falls off above this frequency.

An *audio amplifier* is designed to operate at full gain (that is, with its output within 3 dB of its maximum output) over the range of frequencies that the human ear recognises as sound. This extends from 15 Hz up to 20 kHz.

Note that an audio amplifier is not expected to amplify DC input at full gain.

The high-frequency amplifiers which are the subject of this chapter fall into three categories. *VHF amplifiers* cover the frequencies used for VHF radio transmission, that is from 30 MHz to 300 MHz or even higher. By contrast, the distinctive feature of an *RF amplifier* is its narrow bandwidth. Its centre frequency may be anything from about 30 kHz (low-frequency radio) up to several hundred megahertz (VHF radio). The narrow band is usually obtained by using a tuned circuit, so this type of amplifier is also known as a *tuned amplifier*. The VHF and RF amplifiers operate only at high frequencies but there is a third type of amplifier capable of working at very high frequencies yet also able to amplify DC and low-frequency signals. Its operating range may extend from DC up to about 10 MHz. This extensive range gives the amplifier its name, the *wideband amplifier*. It is also known as a *video amplifier* because amplifiers of this type are often used in video circuits.

Capacitance at high frequency

Capacitors feature in almost all of the amplifiers described in previous chapters, mainly for coupling at input, between stages, and at output. They are also used as by-pass capacitors for stabilising gain. When a capacitor is used for coupling we generally have a configuration that acts as a highpass filter (Fig. 2.13). In a high-frequency amplifier we may reduce the value of the coupling capacitors so as to pass only the high frequencies and to prevent low frequencies from passing through the amplifier. But always there is an upper limit to the frequencies that an amplifier will handle (Fig. 8.1). The reason for this is capacitance occurring within certain of the components, particularly in transistors. The effect of this is negligible in DC and other low-frequency amplifiers but becomes of major importance at high frequencies.

There are two causes of this capacitance. One of these is not really capacitance but has a very similar effect. When a junction is reverse-biased there are minority carriers diffusing across it. The carriers which are on their way through the junction at any instant represent a certain amount of charge 'stored' in the junction. This corresponds to the charge stored on a true capacitor. When the signal level changes there is a change in the amount of charge 'stored' in the junction. But the carriers do not move instantly in or out of the junction. It takes time for this change to take effect, just as it takes time for the charge stored on the plates of a real capacitor to adjust to new signal levels. In this way the minority carriers contribute to the capacitance of the junction.

Apart from this, there is genuine capacitance at the junctions. A BJT can be thought of as three electrodes (collector, base and emitter) separated from each other by pn junctions. There is a depletion region at each junction, which is devoid of charge carriers and is therefore an insulator. The electrodes act like the plates of a capacitor and the depletion region like a dielectric between them. Because the depletion regions are thin the electrodes are in effect very close together and so the capacitance between them is relatively high for the size of the device.

Fig. 8.2 shows the capacitances present in a BJT. The total capacitance varies with the junction, depending on transistor dimensions and operating biases. In an unbiased junction the capacitance is in the region of 2 pF but 'storage' of charge and the reduction of the thickness of the depletion layer in a forward biased junction can raise the effective capacitance to as much as 1 nF.

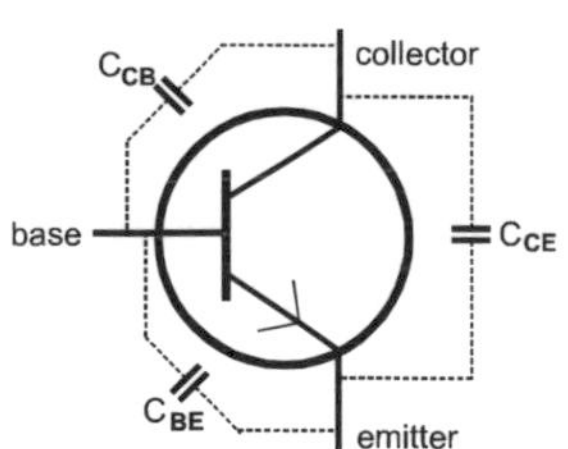

Figure 8.2 *The capacitances in a BJT are represented in this figure by the three capacitors connected by dashed lines. Although they are normally only a few picofarads, they become significant at high frequencies.*

As Fig. 8.2 shows, the base-emitter capacitance acts in a common-emitter amplifier to divert the signal to the emitter and from there to ground. This action depends on frequency because the reactance (the capacitative equivalent of resistance) increases with frequency. We calculate reactance, X_C, of a capacitor using the equation:

$$X_C = \frac{1}{2\pi fC}$$

–3 dB level

We often see the response of an amplifier or a filter quoted 'at the –3 dB level' and may wonder why this particular value is chosen. If you work backward through the decibel equation in the previous box, putting $n = -3$ and calculating x_2/x_1, you find that $x_2/x_1 = 0.7079$. This is the ratio of voltages or currents. Because the power is proportional to the *square* of voltage or current ($P = I^2R$ and $P = V^2/R$ are both derived using Ohm's Law), we square this value to obtain the ratio of the powers of the signals. Doing this, we find that $0.7079^2 = 0.501$. This is close enough to 0.5 so the –3 dB level corresponds to a power reduction of almost exactly half.

For example, at 100 Hz, a 10 pF capacitance has a reactance of 160 MΩ, so the signal is adequately isolated from ground. But at 100 MHz the same capacitance has a reactance of only 160 Ω, and consequently much of the signal is lost to ground. This effect leads to a fall-off in h_{fe} at higher frequencies. There are two critical frequencies:

- the cut-off frequency, at which h_{fe} falls to 3 dB below its maximum level.
- the transition frequency, at which h_{fe} falls to unity.

The transition frequency varies with collector current and is generally highest when the current is a few milliamps or tens of milliamps.

Miller effect

In a common-emitter amplifier and certain other circuits, this is the most important result of capacitance. It occurs within the transistor at high frequencies, at the base-collector junction. On the base side of this junction we have a small and varying base-emitter voltage. On the other side we have the collector-emitter voltage, also varying, but with much greater amplitude than the base-emitter voltage. If the voltage gain of the amplifier is A_V, the voltage on the collector side of the junction varies A_V times as much as the voltage on the base side. The result is that the potential difference across the junction (and so across the base-collector capacitance) varies by $(A_V + 1)$ times as much as the base-emitter voltage. It is as if the base-collector capacitance is $(A_V + 1)$ times its actual value. It may be increased by several hundred times and results in a significant fall-off in gain at high frequencies as more and more of the signal is shunted to the positive rail.

FETs have the equivalent inter-electrode capacitances that BJTs exhibit but they lack the flow of minority carriers through depletion regions, so do not suffer from the effects of charge 'storage'. The Miller effect operates in FETs across the gate-drain junction.

Among the basic types of amplifier, the Miller effect is important only in common-emitter and common-source amplifiers. In the common-collector amplifier, with the collector joined directly to the positive rail, the collector voltage is constant. It does *not* vary by $(A_V + 1)$ times the base voltage so the capacitance at the base-collector junction is only the actual value of, say, 2 or 3 pF. In the common-base amplifier, the emitter (input) and the output (collector) sides are separated from each other by the base layer. This considerably reduces the Miller effect. In addition, the common-base amplifier has low input impedance. Even though this increases at high frequency, it still remains too low to be of concern. For these reasons the common-base connection is favoured for high-frequency amplifiers.

Cascode circuit

The cascode circuit is one of the ways in which we attempt to counter the Miller effect in high-frequency amplifiers. Basically, a BJT cascode circuit comprises two transistors, Q1 in the common-emitter connection and Q2 in the common-base connection (Fig. 8.3). The common-emitter section provides voltage amplification in the usual way. But in a basic common-emitter amplifier the Miller effect causes a roll-off of gain at high frequencies. One way to prevent this occurring is to hold the collector voltage constant, or at least as constant as possible. Ideally we would connect it directly to the positive supply line. But the common-emitter amplifier needs a collector resistor in order to convert the varying collector current into a varying output voltage. The voltage at the collector therefore rises and falls by the full amplitude of the amplified signal, and the Miller effect is unavoidable.

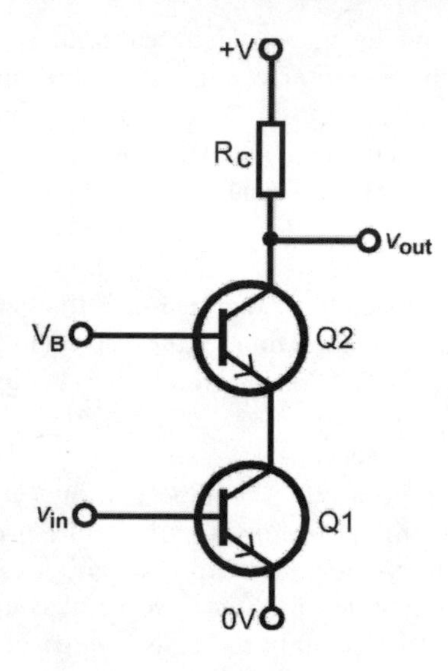

Figure 8.3 *This cascode amplifier is a common-emitter amplifier (Q1) with another transistor (Q2) between the collector of Q1 and its collector resistor R_C.*

To see what can be done about this, look first at a simple common-emitter amplifier on its own (Fig. 8.4). This version has no by-pass capacitor or output capacitor. We measure the output directly from the collector. In the version simulated, the voltage gain was only 2.2, as was predictable from the ratio of the collector and emitter resistors. Fig. 8.5 is a Bode plot of the frequency response of an amplifier of this type. The dashed line represents the –3 dB level, which is generally taken to define the bandwidth of an amplifier. Measurement indicates that the output was above the –3 dB level from 340 Hz to 51 MHz, a bandwidth of 50.7 MHz.

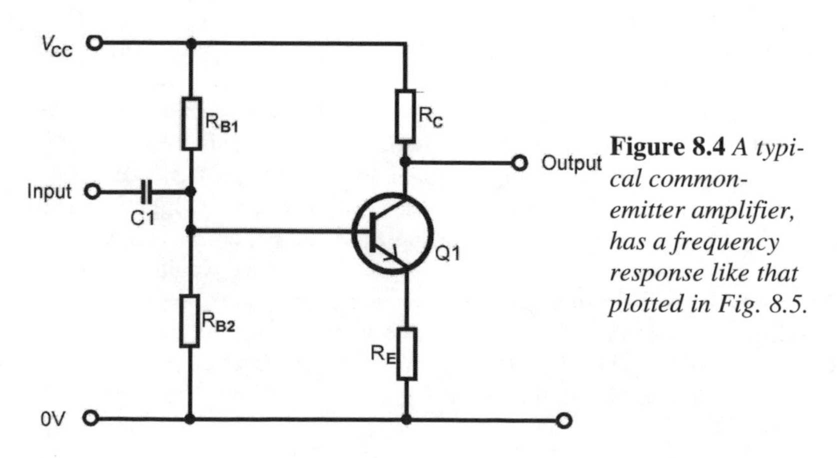

Figure 8.4 *A typical common-emitter amplifier, has a frequency response like that plotted in Fig. 8.5.*

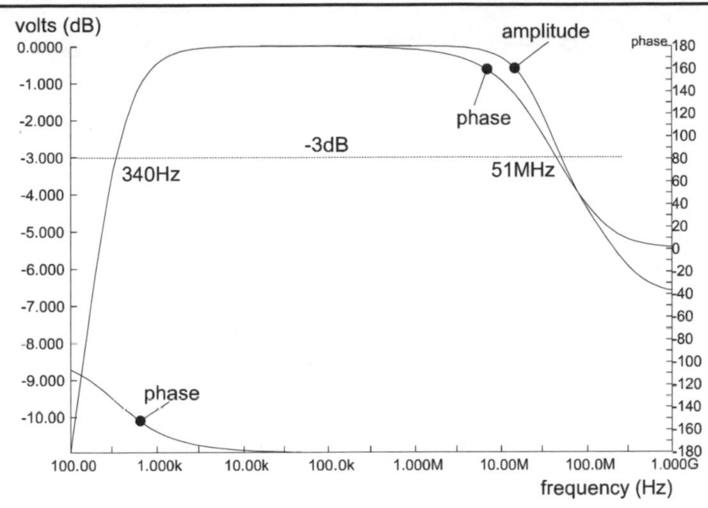

Figure 8.5 *Frequency response of the basic common-emitter amplifier, showing a bandwidth of 50.7 MHz.*

The amplifier is converted into a cascode amplifier simply by inserting a transistor between Q1 and the collector resistor (Fig. 8.6). The base of Q2 is held at a constant voltage V_{BB}. This can be done in various ways, the simplest being to use a pair of resistors as a potential-divider, in the same way as those used to bias the base of Q1. For stability, a large-value capacitor is normally wired between the base of Q2 and the 0 V rail. Output is now taken from the node between the collector of Q2 and the collector resistor. Since the base of Q2 is held at a constant voltage, its emitter must also be at a constant voltage, approximately 0.6 V (V_{BE}) lower. This means that the collector of Q1 is at constant voltage. The collector currents of Q1 and Q2 are at all times equal (ignoring the relatively small addition of the base current of Q2), so R_C passes the collector current of Q1, just as it does in the common-emitter amplifier. Output is as before and gain is 2.2 as before.

The big change is at the upper end of the frequency response (Fig. 8.7). Now the upper –3 dB limit extends as far as 176 MHz. The bandwidth has increased to 175.4 MHz. This is an increase of nearly 3.5 times, all at the higher end, where it is needed. The figure shows a slight hump on the plot at the high-frequency end before it dips down at the highest frequencies. The plot was made with V_{BB} set to 3 V. Slightly decreasing V_{BB} below 3 V

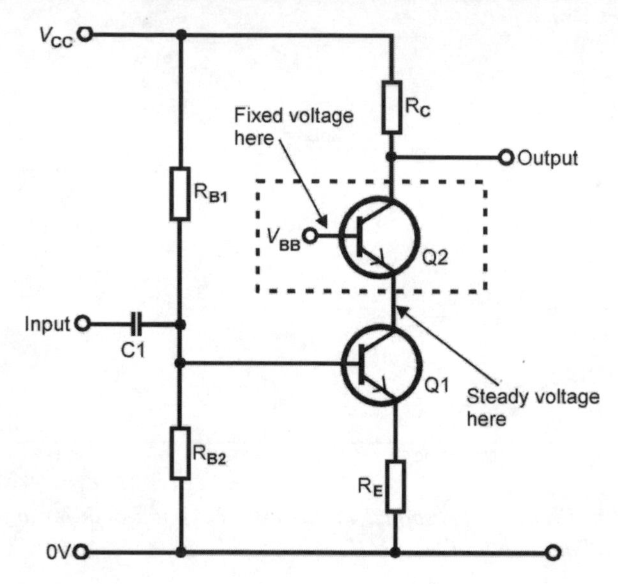

Figure 8.6 *The insertion of a second transistor converts the amplifier into a cascode amplifier, which has the improved high-frequency response plotted in Fig. 8.7.*

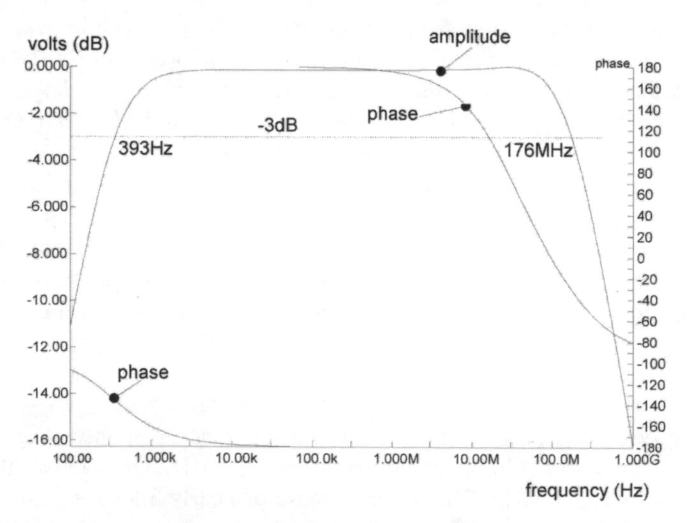

Figure 8.7 *The cascode amplifier has its –3 dB point at 176 MHz, a big improvement on the amplifier of Fig. 8.4. There is a slight peak at about 50 MHz. After that, the amplitude falls sharply to reach –16 dB at 1 GHz.*

accentuates the hump, so increasing the bandwidth further.

Measurements of the voltage at the junction of the collector of Q1 and the emitter of Q2 ascertain that the circuit works as described. The signal is detectable there in inverted form (as expected in a common-emitter amplifier), but the amplitude of the signal is only 140 μV. Contrast this with the 10 mV amplitude of the signal itself and the output amplitude of 20 mV. It is clear that the cascode circuit is an effective way minimising the Miller effect.

Keeping up?

1. What is the difficulty that we usually encounter when we try to amplify a high-frequency signal?
2. List the types of amplifier designed for different frequency ranges.
3. What is a decibel?
4. What is the significance of the –3 dB level?
5. In what two ways may the current gain of a transistor be reduced at high frequency.
6. Explain the Miller effect. In what types of amplifier is it most serious?
7. Define (a) cut-off frequency, and (b) transition frequency.
8. Explain the use of a cascode circuit to eliminate the Miller effect?

Phase

We have used the term 'phase' several times without going into detail. Phase plays a more important part when frequency is high, so we will look more closely at it now. A periodic signal such as a sine wave has a waveform that is repeated regularly, each repetition being known as a *cycle*. The time taken for one cycle is the *period* P of the waveform and the frequency of the waveform is given by $f = 1/P$. These relationships are illustrated in Fig. 1.2. We specify the stage that a signal has reached at any given instant of time as its *phase*, which is expressed as an angle. We take one complete cycle of the waveform to be 360°, beginning from some convenient point such as when the line of the waveform crosses the axis.

Sometimes we express the angle in radians instead of degrees. In this case a complete cycle is 2π rad.

We may also use the term when describing the *phase difference* between two signals of the same frequency. Having the same frequency is essential because two signals with different frequencies would continually be getting out of step and the idea of them having a phase difference is meaningless. Most often we refer to phase difference when comparing the phase of the input to an amplifier with the phase of the output signal. At low frequencies, transistors in the common-collector connection produce an output signal that is *in phase* with the input signal. They go through all stages in their cycle at exactly the same instant. We can say that their phase difference is zero. By contrast, a transistor in the common emitter connection is an inverter. It has a phase difference of exactly half a cycle, or 180°, or π rad. When the input begins to swing positive, the output begins to swing negative. Fig. 8.8 shows this happening in a common-emitter amplifier (the simple one shown in Fig. 8.4) when the signal frequency is 100 kHz. The waveforms for the input signal and the signal after it has passed through the input capacitor C1 to the base of Q1 are exactly equal in amplitude and in phase. The output signal has a greater amplitude and it is 180° out of phase. We can see these facts on the Bode plot of Fig. 8.5. At 100 kHz the amplitude is at its maximum (0 dB) and the phase curve passes through +180°. It also reappears at –180° at the bottom of the plot. This is not a sudden change of phase but is the result of limiting the plotted phase scale from –180° to +180°. Going 180° forward or backward through the cycle brings us to the same place in the cycle.

In Fig. 8.5 the phase difference is reduced when the frequency is greater than about 1 MHz. This commonly happens in amplifiers of all kinds because of capacitance in the transistor and also the Miller effect, if present. We can see the results of this in Fig. 8.9 which is a repeat of Fig. 8.8 but with the signal running at 100 MHz. The input signal and the signal at the base of the transistor are still in phase and of equal amplitude. The input capacitor is passing the full signal, as it did at 100k Hz. But the output signal now leads the input and base signals by only about 36°, as can be confirmed by reference to Fig. 8.5. There has been a change in the timing of the output signal relative to the base signal. This change must have taken place in the transistor, and it due to the effects of capacitance there.
Fig. 8.5 also shows that amplitude and phase alter as frequency falls below

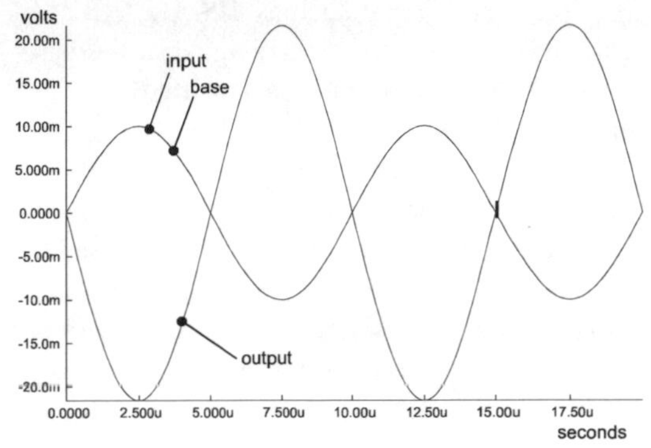

Figure 8.8 *At 100 kHz voltage gain is at its maximum (2.2) and the output is 180 degrees out of phase with the input.*

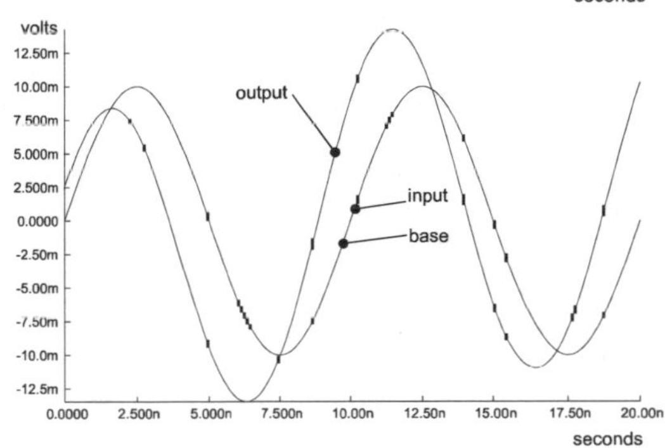

Figure 8.9 *At 100 MHz voltage gain is reduced to about 1.25 and output is 36 degrees in advance of input.*

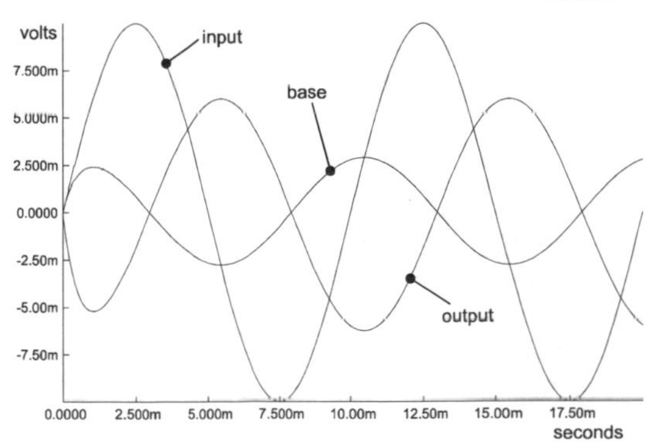

Figure 8.10 *At 100 Hz voltage gain is 0.6. Output lags 108 degrees behind the input but is still 180 degrees out of phase with the signal at the base.*

about 1 kHz. The cause is suggested by Fig. 8.10 in which the frequency has been reduced to 100 Hz. The base and output signals are 180° out of phase, as they were at 100 kHz. Obviously the transistor capacitances are having no significant effect and the transistor is performing correctly as an inverting amplifier. Now the input capacitor is to blame. In conjunction with the biasing resistors it is acting as a highpass filter, and this not only reduces the amplitude of a low-frequency signal such as 100 Hz, but also introduces a phase difference across C1. As a result, the signal at the base of the transistor is about 72 ahead of the input signal. As a result of this, the output is 108° behind the input.

Summing up, the amplifier, designed to operate in the 1 kHz to 1 MHz range has maximum output amplitude and 180° phase difference in the whole of this range. At higher frequencies, amplitude and phase difference are reduced because of capacitance in the transistor and the Miller effect. This is inherent in the transistor, although we may design special circuits such as the cascode to overcome these defects. At lower frequencies, the transistor is unaffected by capacitance but amplitude and phase difference are reduced because of the filtering effect of the input capacitor. Low-frequency response may be improved by using more suitable values for C1 and the biasing resistors.

It is easy to understand why the amplitude of a signal is important, but we might wonder why phase too is important. A 1 kHz sine wave is fed into an inverting amplifier and emerges an instant later. The emerging signal is a half a cycle ahead (or behind) the input signal. At 1 kHz, half a cycle is equivalent to an advance (or delay) of 0.5 ms. In an audio system it might seem that it does not matter if the signal is 0.5 ms early or late in arriving at the speaker. If the phase lead or delay is less than 180°, it should matter even less. But an audio system is carrying sounds of many *different* frequencies simultaneously and it can be seen in the Bode plots (for example, Fig. 8.7) that phase delay depends on frequency. Within the working range of the amplifier we should aim for all frequencies to have the *same* phase delay. In Fig. 8.7, this applies to all frequencies between about 10 kHz and 1 MHz. The phase response of this amplifier at lower frequencies makes it unsuitable as an audio amplifier because signals of different audio frequencies are delayed (or advanced) by differing amounts. On arriving at the output they are jumbled out of order, resulting in distortion. There is more about this in Chapter 9 under the heading of *Group delay distortion*.

Another cascode amplifier

The cascode amplifier of Fig. 8.11 operates on a different principle from that in Fig. 8.5. There we used a common-emitter amplifier to provide gain,

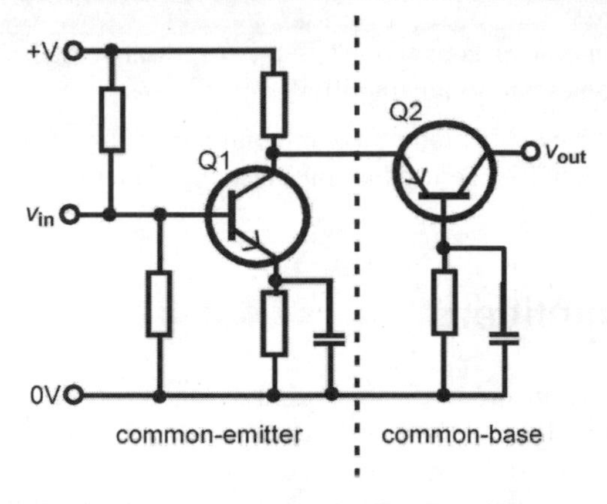

Figure 8.11 *There is minimal Miller effect in the common-emitter stage because it has low gain. There is no Miller effect in the common-base stage.*

and used a common-base amplifier to hold the collector voltage constant and so avoid the Miller effect. In Fig. 8.11 we again have common-emitter and common-base amplifiers but their functions are different. The gain is provided by the common-base amplifier, which is based on a pnp transistor Q2. This is feasible because common-base amplifiers have much the same voltage gain as common-emitter amplifiers. Because the emitter and collector of Q2 are separated by the base, the Miller effect is absent. The common-base amplifier on its own would serve very well except that it has a low input resistance. This is where the common-emitter amplifier is important, for this has high input resistance if its biasing resistors are of high value. But common-emitter amplifiers are subject to the Miller effect so we have apparently re-introduced this effect into the circuit. As we described earlier, the Miller effect arises when there is capacitative coupling between the base current and the much larger collector current. In this common-emitter stage we chose the values of the resistors and by-pass capacitor so that gain is very low. Then the collector current is *not* much larger than the base current, and the Miller effect is minimised.

Keeping up?

9. What is meant by phase?

10. What is the phase difference between the input and the output of a BJT

in the common-emitter connection? Is this necessarily the same when the transistor becomes part of an amplifier?

11. How can the combination of a common-emitter amplifier followed by a common-base amplifier be used to minimise the Miller effect?

Tuned amplifiers

A tuned amplifier incorporates some kind of tuned network. Usually this is an inductor-capacitor (LC) network as pictured in Fig. 8.12. The network replaces the drain resistor of the common-source amplifier. The principle is applicable to BJT amplifiers too, such as the common-emitter amplifier, where the network replaces the collector resistor.

In the network of Fig. 8.12 the reactance of the capacitor is high at low frequencies ($X_C = 1/2\pi fC$) and that of the inductor is low ($X_L = 2\pi fL$). A high reactance in parallel with a low reactance is equivalent to a slightly lower reactance, so the overall reactance of the network is low and the signal at the collector is by-passed to the positive rail. As signal frequency is increased the reactance of the capacitor decreases and that of the inductor increases. One particularly interesting frequency is given by:

$$f = \frac{1}{2\pi\sqrt{LC}}$$

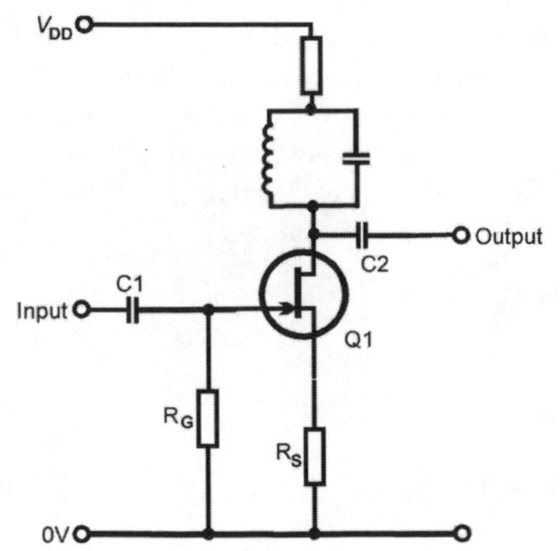

Figure 8.12 *A common-source amplifier is tuned to a range of high frequencies by including an LC network in the drain circuit. A resistor acts to dampen the response to provide a broader frequency range.*

At this frequency the reactances of the capacitor and inductor are equal and the total parallel reactance is a maximum. We choose the values of L and C so that this maximum occurs at the intended operating frequency of the amplifier. If frequency is increased beyond this point, the reactance of the inductor increases further but the capacitor has low reactance and the signal is lost to the positive rail. If the frequency decreases, the reactance of the inductor decreases and the signal is again lost to the positive rail. At the tuned frequency the network resonates. Energy is alternately stored in the electrical field in the capacitor and in the magnetic field in the inductor and it requires very little energy from the main circuit to maintain the oscillations. At this frequency the network acts as a pure resistance of large value and the signal developed across it is at a maximum. The result is that the amplifier has high gain at its operating frequency but low gain at lower and higher frequencies.

Fig. 8.13 demonstrates what happens. There are 10 curves in this figure, additionally illustrating the effect of the resistor in series with the LC network. When the resistor has low value, the network resonates vigorously and the response peaks sharply. This is undesirable, possibly leading to instability. The amplifier becomes an oscillator. With a resistor of higher value more of the energy in the network is dissipated so that, although there

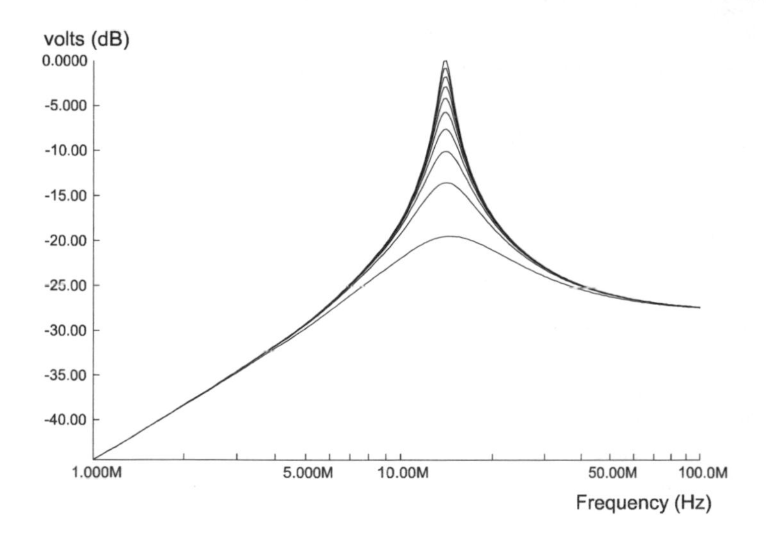

Figure 8.13 *Varying the damping resistor from 1 kΩ (highest curve) to 10 kΩ (lowest curve) broadens the bandwidth of the amplifier.*

is still a peak in the response, it is much broader. As we increase the resistor, the effect of variations in the reactance of the LC network itself become relatively less important, so the change of output amplitude with frequency is less marked.

Ideally, a radio-frequency amplifier would amplify equally all frequencies within a given bandwidth, and none of those outside it. In practice, this is hard to achieve. Either the band is broad (like the lower curve in Fig. 8.13) so that frequencies immediately outside the desired band are amplified almost as much as those inside it. Or the band is very narrow (like the upper curves) so that the sidebands of a transmission are inadequately amplified. In the first case, we say that Q is too low, and in the second case, Q is too high.

None of the family of curves in Fig. 8.13 is ideal. We need a curve that is broader yet with sharply dropping, almost vertical, sides. This is difficult to obtain from an amplifier containing a single transistor. But it is possible to obtain the required effect by using a multistage amplifier, each stage having a moderate Q. Although a multistage amplifier may be used more often to obtain increased gain, it is quite often used to obtain good selectivity as well. An amplifier that operates at only one frequency or over a very narrow range of frequencies has the advantage that signals outside this range are either strongly attenuated or eliminated altogether. These include

Q factor

The *quality factor* of a circuit is generally known as 'Q' for short. It describes the selectivity of a circuit. For example, a radio receiver with low Q has low selectivity and receives transmissions from several stations simultaneously. To tune to a single station when there are other stations transmitting on nearby wavelengths, we need a receiver with high Q.

In one definition, $Q = B/f_o$, where B is the bandwidth and f_o is the centre frequency (generally the resonant frequency of the LC network). It can also be defined in terms of the inductance, damping resistance, and centre frequency, $Q = 2\pi f_o L/R$. Since f_o depends on both L and C, the expression takes C into account. Q is high when L is large or R is small.

noise and unwanted signals such as mains-induced hum, or harmonics of the primary signal. Another bonus is that stray capacitances and capacitances within the transistor itself, usually a cause of signal attenuation at high frequencies, have now been 'taken over' and are included as part of the capacitance of the network. Instead of being a nuisance they now contribute to the intended behaviour of the circuit.

High-frequency devices

The results described so far in this chapter have been obtained from circuits employing small-signal transistors such as the BC548. These give satisfactory results in many instances but we can obtain better high-frequency performance by using specially-designed high-frequency transistors. An example is the ZTX325, with a transition frequency of 1.3 GHz, compared with an f_T of 300 MHz for the BC548. To achieve this kind of performance the transistor is built to have a base-emitter junction capacitance of only 1.18 pF compared with a value of 13 pF for the BC548. Similarly, its collector junction capacitance is 1.54 pF compared with 4 pF. With such small capacitances it is to be expected that all capacitative effects will be much less with the ZTX325 and thus high-frequency performance will be correspondingly greater.

To investigate this, the BP548 transistors in the circuit of Fig. 8.6 were replaced with a pair of ZTX325 transistors. The Bode plot is similar to that of Fig. 8.7 except that the upper –3 dB point occurs at 900 MHz. Using radio-frequency transistors multiplies the bandwidth a further five times.

Differential amplifiers

The basic differential amplifier and some of its variants are described in Chapter 6. These amplifiers have a very good high-frequency response. Taking the single-ended amplifier of Fig. 6.10 as an example (redrawn in Fig. 8.14) we see that the collector of Q1 is connected directly to the positive supply rail. This holds the collector at a constant voltage so that the Miller effect can not occur. The other transistor is connected as a common-base amplifier, which is not subject to the Miller effect because its base is connected directly to ground. Thus the differential amplifier has an excellent high frequency response.

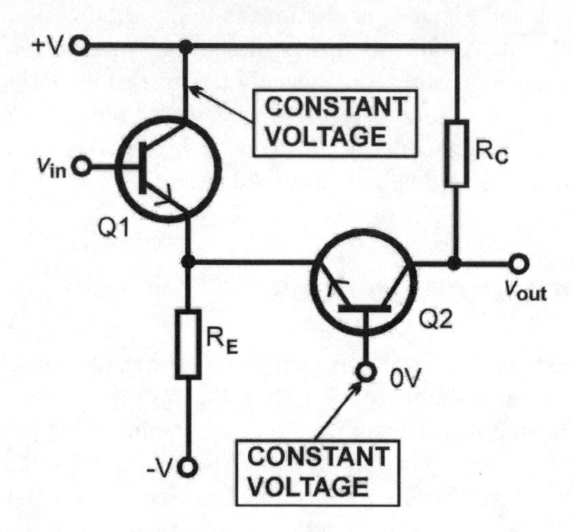

Figure 8.14 *In the single-ended differential amplifier, the constant voltages prevent the Miller effect from occuring and give good high-frequency response.*

The freedom from Miller effect of the single-ended amplifier depends on the collector of Q1 and the base of Q2 being connected directly to fixed voltages. If we want to build a differential amplifier with symmetrical outputs, Q1 needs a collector resistor. This brings back the Miller effect and another approach must be tried. Fig. 8.15 has the transistors of the basic differential amplifier replaced by cascode amplifiers. As explained above, cascode amplifiers have a good high frequency response, so they improve the performance of the differential amplifier.

The two R_E resistors shown in Fig. 8.15 are optional. They can be omitted altogether (as in Fig. 6.6). If present, we may use two fixed resistors or the two sections of a variable resistor (as in Fig. 6.9), to allow the circuit to be balanced.

Emitter resistors provide *current feedback* which helps to stabilise the amplifier. If for any reason (for example, rise in temperature) the collector current begins to rise, the rise in current through the emitter resistor increases too, raising the voltage across R_E. This raises the voltage at the emitter terminal of the transistor, and slightly decreases the base-emitter voltage. The reduction in V_{BE} causes a fall in base current and hence a fall in collector current. The feedback has a negative effect, serving to keep the

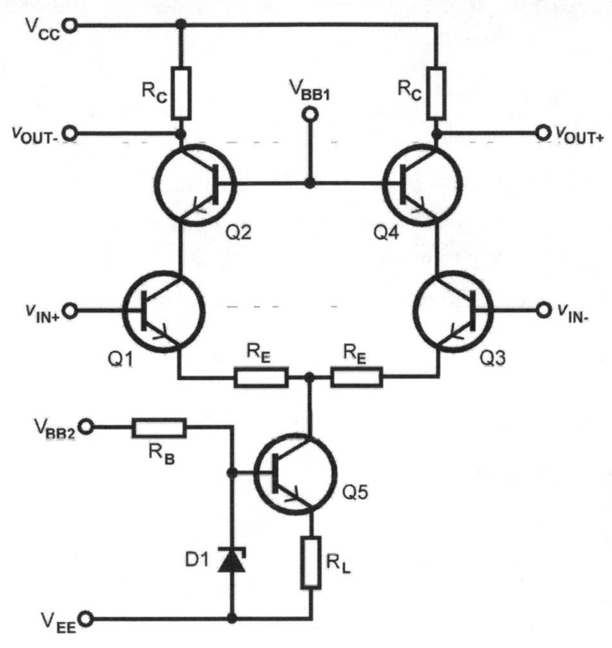

Figure 8.15 *This differential amplifier, has cascode amplifiers at the inputs to reduce Miller effect. This ensures a good high-frequency response.*

amplifier stable. The action is the same as that of the source resistor in Fig. 2.16.

Inevitably, the use of negative feedback reduces voltage gain. As explained in Chapter 5, the voltage gain of a common-emitter amplifier equals $-R_C/R_E$. With no emitter resistor, only the emitter resistance of the transistor need be taken into account. This is low, so gain is high. But the amplifier of Fig. 8.15 typically has a resistor of about 1 kΩ in series with this, so voltage gain is limited to 10 or 20 times. Voltage gain may also be limited by the need for low output resistance. As explained in Chapter 6, both gain and output resistance depend on the value of R_C, so low output resistance and low gain go together. One solution is to cascade several low gain differen tial amplifiers. The problem of incrementing the quiescent output voltage at each stage is met by using amplifiers with a pnp output stage (Fig. 6.15). This too is a type of cascode circuit so it has a good high-frequency response.

Keeping up?

12. Describe the action of the inductor-capacitor network in a tuned amplifier.

13. What is meant by Q? Why is a high Q not ideal for a radio amplifier?

14. Name a radio-frequency transistor type. How does it differ from an ordinary low-frequency transistor?

15. Why is a single-ended differential amplifier relatively free of the Miller effect?

Voltage followers

A BJT common-collector (emitter follower) amplifier (Fig. 6.1) behaves well at high frequencies. This is because the collector voltage is held constant at the positive supply voltage and the Miller effect can not occur. The main disadvantage of this amplifier is that, although it is called a 'follower', the output voltage is always about 0.6 V less than the input voltage. This 0.6 V offset, due to V_{BE}, varies with temperature. For a change of temperature of one degree (Celsius or kelvin) the offset decreases by 2.1 mV. A transistor may easily experience a temperature change of 100 degrees as equipment warms up, producing a 0.2 V decrease in offset.

A JFET common-drain (source follower) amplifier does not have these problems. The amplifier in Fig. 8.16 has a second JFET acting as a constant current drain. The constant current is programmed by the value of the two equal resistors R1 and R2. The transistors are identical, being formed on the same chip. The same current flows through both resistors.

To understand the circuit, begin by looking at Q2, the constant current source (see box). The current I_D through this is set by the value of R2. The same current flows through R1, so Q1 is operating with voltage *differences* between its terminals identical to those in the Q2/R2 sub-circuit. Whatever the value of v_{IN}, the gate voltage and the voltage at the 'lower' end of the resistor are equal. They are equal in Q2 because they are wired together. They are equal in Q1 because of the operating conditions of the transistor, as set by the constant current. Therefore v_{OUT} is equal to v_{IN}.

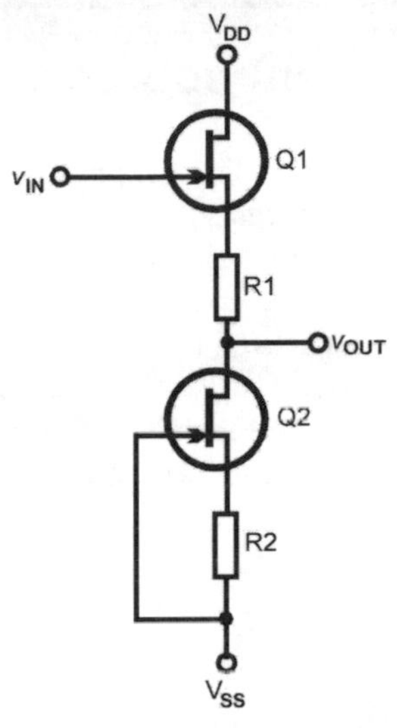

Figure 8.16 *A source follower with a constant current drain has the advantage of zero output offset voltage.*

This circuit has a high bandwidth, and its output follows the input very accurately. The only conditions for this are that the input signal is small and that only a small current is drawn from the output. One snag is that when it is connected to a load, there is usually stray capacitance due to the load between the output terminal and the 0 V line. The capacitance can be charged quickly by current flowing from Q1 which has low output resistance. But it is discharged only slowly through R2. This introduces an asymmetrical distortion into the signal when amplitude is large.

Neutralization

Capacitance between the layers of a transistor or between the electrodes of a valve produces a coupling between input and output stages. Often the coupling provides positive feedback and it is likely to result in oscillation or at least in severe distortion. In extreme cases it converts the amplifier into an oscillator. To avoid this happening it is necessary to cancel the capacitance by techniques known under the general heading of *neutralization*. The usual approach is to make use of capacitative negative feedback. In valves, a similar tendency to oscillate may result from inductive coupling between the screen grid and the cathode of tetrodes. Cancelling this kind of coupling is another instance of neutralization.

Fig. 8.18 is a radio-frequency common-emitter amplifier with a inductor-

JFET constant current source

The basic circuit consists of a JFET with its gate connected to its source (Fig. 8.17). With the gate-source voltage V_{GS} thus made equal to zero, the current through the transistor takes a fixed value known as I_{DSS}, which is independent of the voltage. The conductive channel in the JFET is pinched off at one end. An increase in voltage tending to cause an increase in current is compensated for by an increase in the proportion of the channel which becomes pinched off, reducing current flow. This is why the device may be used to supply (or sink) constant current.

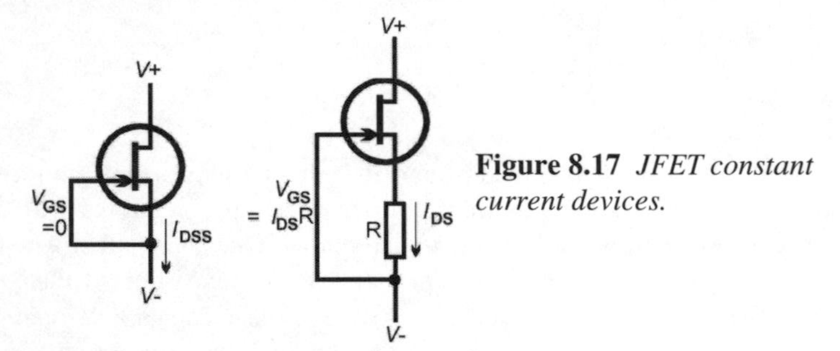

Figure 8.17 *JFET constant current devices.*

If the gate is made negative of the drain, pinch-off occurs at lower gate-source voltage and the constant current has a lower value. V_{GS} is set by wiring a resistor between gate and source. The constant current flows through the resistor, creating a constant V_{GS}, with the gate negative of the source. In turn, this negative V_{GS} results in a constant current of a specific value I_D, smaller than I_{DSS}.

capacitor tuned circuit between the collector and the positive supply. Output is taken inductively from a coil L2 wound on the same former. Normally the tuned circuit would be connected to the supply at the junction of the capacitor and inductor (at point A) and the capacitor C_N would not be there. In this circuit the positive connection is a tap exactly half-way along the inductor. The effect of this is to cause the signals at A and B to be exactly equal in amplitude but opposite in phase. In the absence of C_N there is positive feedback through the collector-base capacitance of Q1, drawn in the figure as an imaginary component with dashed connecting lines. When we wire the neutralizing capacitor C_N into the circuit, the signal from point

A is fed back. Being opposite in phase to the feedback signal through C_{CB} the signal coming through C_N rates as negative feedback. Both feedback signals are fed to the input line, where the negative feedback cancels out the positive feedback. In practice the signals may not be exactly 180° out of phase and a resistor may be wired in series with C_N to adjust the phase difference.

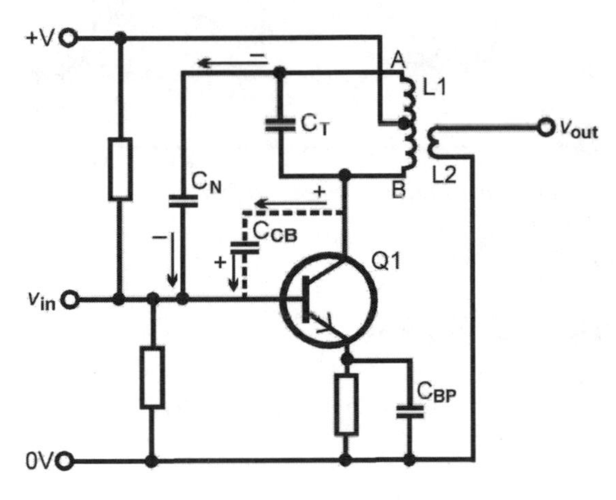

Figure 8.18 *The positive feedback through C_{CB} is neutralized by the negative feed backthrough C_N.*

It is often very easy to neutralize a push-pull circuit because of the symmetry inherent in its operation. Fig. 8.19 shows an example of neutralizing a triode push-pull amplifier. The couplings to the previous and following stages are not shown in the drawing, but there are coils wound on the same formers as the inductors in the two tuned circuits. The main capacitance likely to cause oscillation is that which couples output back to input, in this case the capacitance between the grid and plate (C_{GP}).

In a push-pull amplifier, the two valves operate in anti-phase so it is easy to neutralize the circuit by cross-connecting a pair of capacitors C_N. The feedback from one side of the amplifier has the phase required to neutralize the capacitance effect on the other side of the amplifier. Only a small capacitance is required. Sometimes it is sufficient to lead a wire from the grid terminal of one valve and run it close to the plate terminal of the other valve. Capacitance is adjusted by varying the distance between wire and plate terminal.

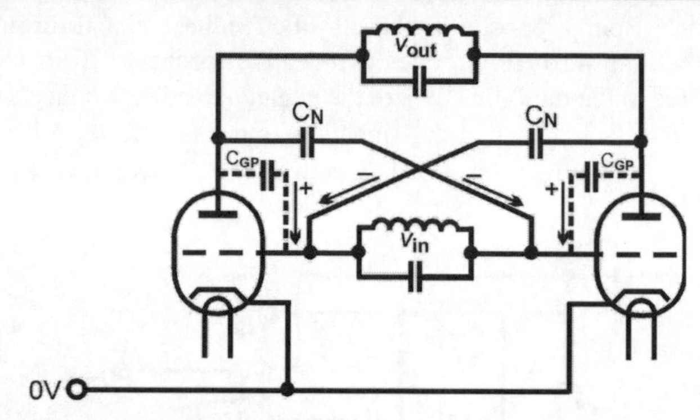

Figure 8.19 *Cross-connected capacitors provide neutralization in a push-pull valve amplifier.*

Summing up

The gain of an amplifier falls off at high frequencies mainly because of the effects of capacitance.

This comprises true capacitance between the electrodes of the transistors, and the effects of 'storage' of charge on its way through the transistor.

The Miller effect occurs when voltage gain in a transistor causes the normal capacitance between layers to be greatly increased. It does not occur in voltage-followers.

Measures to avoid the effects of capacitance, particularly the Miller effect, include:

- cascode circuits which hold the collector at a constant voltage,
- using a common-base amplifier, which is immune to the Miller effect,
- specially constructed high-frequency transistors with low capacitance,
- using single-ended differential amplifiers,
- neutralization.

The stage in its cycle that a periodic signal has reached is called its *phase*.

Input and output signals may not be at the same stage on their cycle and then are *out of phase*.

The phase difference between signals of the same frequency is expressed in degrees (from 0° to 360°) or in radians (from 0 to 2π).
An amplifier may exhibit no phase difference between input and output (non-inverting amplifier) or a difference of 180° (inverting amplifier). These differences result from the way the transistors are connected.

Phase differences of other amounts, varying with frequency, may be caused by capacitance effects in the transistors or in the inter-stage coupling capacitors or inductors.

Test yourself

1. Explain the reasons for the shape of the frequency response of a common-emitter amplifier, as illustrated by the Bode plot of Fig. 8.1.

2. Compare and contrast the cascode amplifiers illustrated in Figs. 8.3 and 8.11.

3. What is neutralization? Explain it by reference to one or more circuits.

4. Why are follower amplifiers not affected by the Miller effect?

5. Describe how a radio-frequency tuned amplifier works. Why is it important to dampen the resonance of the network by wiring a resistor in series with it?

9
Imperfections

In the ideal amplifier, the output voltage, current or power is at every instant constantly proportional to the input voltage, current or power. In other words, the gain of the amplifier is constant. We listed this criterion for the ideal amplifier along with others at the end of Chapter 1. In this chapter we shall look at some of these criteria in more detail and discuss various reasons why they may not be met and what we can do (if anything) to make a real amplifier behave in a nearly-ideal way.

An amplifier which has constant gain (whether of voltage, current or power) is said to be *linear*. A plot of voltage input against voltage output is a *straight line.* Linearity is not an all-or-nothing characteristic. Most amplifiers are linear over at least a part of their operating range, though for many of them linearity ceases at higher frequencies. In many applications this may not matter but amplifiers intended to operate linearly at high frequencies need special design, as described in Chapter 8.

Harmonic distortion

Linearity determines the maximum amplitude that the amplifier can deliver. Signals of small amplitude may be amplified in a perfectly linear way. Signals of larger amplitude may force the amplifier into the range in which gain begins to fall off. The extreme excursions of the signal are clipped, causing distortion of the waveform. Distortion of the waveform produces a series of *harmonics*, which are signals of higher multiples of the fundamental frequency. Attempting to avoid this, we usually design an amplifier so that its quiescent (no-signal) output voltage is close to half the supply voltage. Then the output has freedom to swing equally on either side of its quiescent level without approaching the power-line voltages too closely. Even so, because the transconductance of BJT varies with collector current and that of an FET varies with drain current, there is always a certain amount of harmonic distortion at any signal amplitude. The higher harmonics can be removed by filtering or, in an audio amplifier, they may be of too high a frequency to be heard. But some of the power of the input signal has been converted to signals that have to be removed or are undetectable, so power is wasted. Harmonic distortion is lower in the Class B dual npn amplifier which has a centre-tapped output transformer (Fig. 7.9). This is because the currents due to the even-order harmonics ($2f$, $4f$ and so on) flow in the primary coil of the transformer in opposite directions. They tend to cancel out. Odd order harmonics do not cancel in this way but the reduction of second order harmonics contributes significantly to a reduction in distortion.

Reasonable linearity is essential for an amplifier in which the amplitude of the signal is one of its key features. Linear amplification is important for speech and musical signals, for example, though a considerable amount of distortion is acceptable in some applications, such as low-cost intercom systems. In the field of amplitude-modulated radio, the amplitude *is* the signal so the need for linearity is paramount. By contrast, in frequency-modulated radio, changes of amplitude are immaterial and a linear amplifier is not required. A non-linear amplifier can be used, giving the designer the opportunity to optimise some other feature, such as frequency response and particularly high power output.

Nowadays, high-power MOSFETs (for example, VMOS) are readily available. Their high input impedance and very low output impedance are important in many applications. In addition, their response is more linear than that of BJTs. The more linear response, the less the harmonic and intermodulation distortion.

Modulation

When signal A, usually of higher frequency, has one or more of its characteristics altered by another signal B, usually of lower frequency, signal A is said to be *modulated* by signal B. In radio, and several other applications, A is referred to as the *carrier signal*. Modulation may take various forms, including:

Amplitude modulation (AM).
Frequency modulation (FM), as in Fig. 1.3.
Phase modulation (PM), which is similar to FM but varies phase.

Intermodulation distortion

If two or more signals of different frequencies are fed into an amplifier simultaneously, and if the amplifier is not absolutely linear, then a number of spurious frequencies appear in the output. These consists of 'sum and difference' frequencies, and the further the amplifier departs from linearity the greater their magnitude. If the frequencies are f_1 and f_2 the output contains signals at $f_1 + f_2$ and $f_1 - f_2$. These are second order intermodulation signals. Third order signals have frequencies f_1+2f_2, f_2+2f_1, f_1-2f_2, f_2-2f_1, $2f_1-f_2$ and $2f_2-f_1$. Fourth order signals have frequencies of the form f_1+3f_2, $2f_1+2f_2$ and so on. Fifth order signals have frequencies in the form $3f_1+2f_2$. Even order intermodulation signals have frequencies that differ widely from the two originals, so they are outside the working range of the amplifier and can be filtered out. This leaves only the odd order signals to deal with. Some of these may have out-of-range frequencies, while others have negative frequencies. Given that the coefficients of f_1 and f_2 are m and n respectively, the most important intermodulation products are those in which either m or n is negative, in which m and n differ by 1, and the order is odd. In practice the most important frequencies are the third-order products, $2f_1-f_2$ and $2f_2-f_1$. These have the greatest amplitude. They lie closest in frequency to the original signals and are therefore difficult or even impossible to filter out.

The above is an account of what happens when only two frequencies are fed to the amplifier. Typically there are far more. In addition, non-linearity generates harmonics of the original signals and, if strong enough, these may generate significant intermodulation products. It is clear that when we wish to be as free as possible of spurious frequencies it is important for the amplifier to be as linear as we can make it. In general, Class A amplifiers are less subject to intermodulation distortion than Class AB or Class B.

In radio amplifiers intermodulation occurs between carrier frequencies, being received at the same time. The same occurs in telecommunications systems, where channels are stacked by frequency and are all passed through the same amplifier. Intermodulation can occur between the carriers of any pair of channels.

When a radio frequency carrier is being amplitude modulated by a signal (baseband) there is intermodulation between the carrier and the signal it is carrying. Given speech or musical signals with frequencies ranging from about 20 Hz to 20 kHz, each frequency present produces sum and difference frequencies, so that the spectrum of the output from the amplifier or transmitter consists of the carrier frequency surrounded at a short distance on either side by two sidebands. The sidebands range from 20 Hz to 20 kHz below and above the carrier frequency. This is not essentially an instance of distortion, for the sidebands contain all the information necessary to reconstruct the original transmitted signal in the receiver. The fact that both sidebands contain the same signal information and that the carrier contains none means that it is wasteful of power to transmit the whole signal. In single-sideband (SSB) radio transmitters the carrier frequency and one of the sidebands are suppressed, leaving only one sideband to be transmitted.

Group delay distortion

An amplifier contains resistors, capacitors and sometimes inductors, so it is inevitable that it acts as a filter. The effect of a filter is to attenuate certain frequencies. In most amplifiers, signals of the lowest and highest frequencies are the most attenuated, so that amplifier is behaving as a bandpass filter. This is normal and usually acceptable. We know the extent of the passband and confine our use of the amplifier to frequencies within that band. But attenuation is not the only thing that happens to the signal as it passes through the amplifier. As explained in Chapter 8, capacitance and inductance both affect the phase of a signal and this effect is frequency-dependent.

Fig. 9.1 shows the phase response of a multistage BJT amplifier of the type illustrated in Fig. 6.16. At frequencies greater than about 1 kHz the output is virtually in phase with the input. This is as expected, since the amplifier consists of two inverting stages. Performance is less satisfactory at low frequencies. With falling frequency the output leads the input by increasing amounts, until at 7.2 Hz it leads by 180°. Further drop in frequency results in increasing phase lead (the phase curve reappears at the bottom of the graph), reaching about 210 ° at 0.5 Hz.

The other curve in Fig. 9.1 shows *group delay*. This is defined as the negative of the rate of change of phase with respect to frequency. In other words, it is the *gradient* of the phase curve of Fig. 9.1, but regarded as positive if phase decreases with frequency. Ideally, phase does not change at all with frequency, so the group delay is zero. More often there is a small

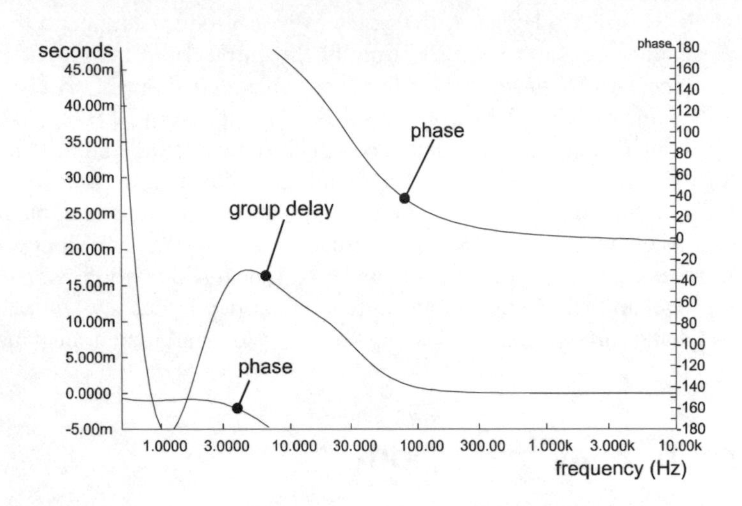

Figure 9.1 *The phase response below 300 Hz shows a relatively large change of phase with change of frequency. Consequently there is a significant rise in the group delay.*

change of phase with change of frequency. The phase curve above about 100 Hz in the figure is like this. The gradient of the curve is small and therefore group delay is small too. Distortion is low. The phase curve is steep between 5 Hz and 300 Hz, so group delay is high, up to 15 ms. Signals of different though fairly close frequencies in the range 5 Hz to 300 Hz are delayed by widely differing amounts so that the signal emerging from the amplifier is badly distorted. It is rather more difficult to interpret the graphs for lower frequencies because the frequency scale is logarithmic.

Keeping up?

1. What is meant by a linear amplifier?

2. The higher harmonics of a signal may have too high a frequency to be significant, but it is still a good idea to try to filter them out. Why?

3. What is intermodulation distortion and which are usually its two most important components?

4. What is group delay and why is it important for it to be low?

Noise

Noise is defined as any unwanted signal superimposed on the signal that is being or has been amplified. This can include a type of noise often referred to as *interference.* This is a signal that has found its way into the amplifier circuit from an external source. It may be a radio signal from a nearby powerful transmitter or it may be a 50 Hz hum due to electromagnetic interference from nearby mains cables or mains-powered equipment. Other external sources of interference are spikes on the mains supply due to switching of inductive loads, such as refrigerator motors, or electromagnetic fields due to lightning strikes. If these unwanted signals get into the amplifier circuit, particularly if they are present in the early stages and are subsequently amplified along with the wanted signal, they can be a great nuisance to the user. If the noise has greater amplitude than the wanted signal it is usually impossible to eliminate it. It is sometimes possible to filter out signals of a specific frequency (such as mains hum) by including a notch filter in the circuit, but other types of signal, once present, can not be got rid of. It is essential to prevent them from ever getting into the amplifier circuit.

If the interference is mains-borne, it may be prevented from reaching the amplifier circuit by using line filters and transient suppressors. If the interference is electromagnetic (which includes interference generated by lightning, and by the sparking of vehicle ignition systems, heavy-duty switches and welding equipment) it may be minimised by screening the circuit in an earthed metal enclosure and by screening all input leads. Cable screening consists of copper braiding, one end of which is connected to the earthed enclosure. Radio frequency interference may enter a circuit through one of the input lines, or maybe along the power lines from another stage of the circuit which is operating at high frequency. In Fig. 9.2, a DC current is flowing in the power line, which includes a coil wound on a metal former. If the DC has radio frequency interference superimposed on it, the rapidly-changing RF component is opposed by currents generated in the coil because of Lenz's Law. This states that if a current is flowing through

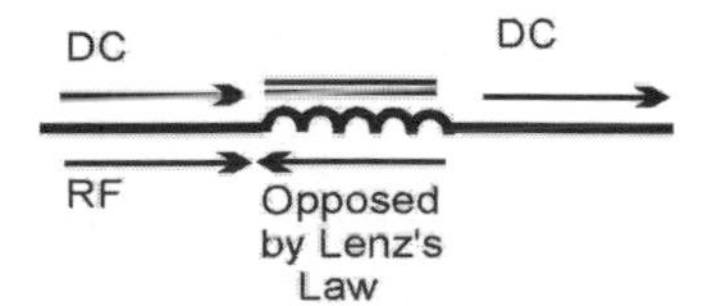

Figure 9.2 *A choke on a power line lets DC and audio frequencies through but eliminates RF.*

a coil and there is an attempt to *change* that current (make it bigger, or smaller, or turn it off) a magnetic field develops in the coil to *oppose* the change. The higher the frequency, the more rapid the change, the stronger the opposition. The effect is to suppress or *choke* the RF while leaving the DC unaffected. Instead of a choke coil we may thread the lines through ferrite beads which have the a same action.

When wiring up an amplifier or connecting amplifiers to other equipment to build up a system, we must take care to return all 0 V lines to a single point in the system. If this is not done, we may unwittingly create loops in the 0 V rail. If electromagnetic fields are present, they will generate currents in the loops and these may find their way into the amplifier, causing interference.

Electrical noise

This is noise that originates within the circuit and is a direct result of the physical carrying of electrical charge. Although it is relatively small, any noise that is generated in the early stages of an amplifier is amplified along with the wanted signal and may become a serious problem by the time it reaches the output. It is not possible to eliminate this kind of noise because it is inherent in the mechanisms of electrical conduction, but it is possible to minimise its effects by suitable circuit design. There are three main types of electrical noise, Johnson noise, shot noise and flicker noise.

Johnson noise or *thermal noise* is due to the random motion of charge carriers (usually electrons) when they are excited by thermal energy. They produce random changes of *voltage* in the conductors. In an audio amplifier it may be heard as a hiss when there is no other signal present. The hiss has no particular frequency and we refer to it as *white noise*. The sounds of escaping steam or of waves on the beach draining through shingle are similar and for the same reason, for these too are the result of randomly moving particles. Rather than say that Johnson noise has no particular frequency, we could say that it includes all frequencies. Noise occurs equally at all frequencies within the bandwidth of an amplifier, so the wider the bandwidth the more noise is generated.

Johnson noise is generated whenever a current passes through a resistive material, and since all components in a circuit have a greater or lesser amount of resistance, the noise is generated in components of all types. The greater the resistance, the greater the noise. To be more precise, the noise level is proportional to the square root of the resistance. Noise is also proportional to temperature, actually to the square root of the absolute temperature. Noise in circuits may be minimised by keeping resistance low (not always an easy thing to do) and by keeping the circuit cool.

Shot noise When a current is flowing, the number of electrons moving past a given point in a period of time averages out at a fairly constant number. But at any particular instant there may be more electrons or fewer electrons than average. There are random changes in the *current* (compare with Johnson noise, which is a voltage noise). The current varies slightly above and below its average value. These irregularities in the current act like a signal and, in an audio amplifier, are heard as white noise. Shot noise depends on the size of the current. The larger the current, the less the effect of individual electrons, and shot noise is small. But with currents in the picoamp range, individual electrons play a relatively large part and shot noise is important.

Shot noise is greatest in semiconductor devices because these depend on the random diffusion of individual electrons. In metallic conductors the electrons move as a group under the influence of the electric field (we call this *mass flow*) and shot noise is very much less.

Flicker noise is also known as *1/f noise* because its amplitude is inversely proportional to frequency. Like shot noise it consists of random variations in the flow of *current*, and it is most significant in semiconductors. Its cause is unknown but it thought to be due to irregularities in semiconductor materials. It also occurs in thermionic valves, probably because irregularities in the surface of the cathode cause the emission of electrons to be uneven. Flicker noise is so small at frequencies above 1 kHz that it can be ignored. But it is important in low-frequency (100 Hz down to DC) amplifiers.

From the above descriptions we see that Johnson noise originates as a voltage noise while the other types of noise originate as current noises. In practice, circuits include resistance which converts current noise into voltage noise, so it is often convenient to consider all forms of noise as voltage noise.

The *signal to noise ratio* is a useful way of expressing the amount of noise present in a system. It is practical measure because we can normally tolerate more noise in a system in which the signal is large than in a system in which the signal is small. Being a ratio, the unit of SNR is the decibel and:

$$SNR = 10 \log_{10}(v_s^2/v_n^2)$$

We use the *squares* of the rms voltages of signal (v_s) and noise (v_n) because the SNR is a ratio of the signal and noise *powers*, and power is proportional to the square of the voltage.

Root mean square

This is a useful way of expressing the average voltage of a signal. If we quote its amplitude (as we may do sometimes when calculating gain) we are looking at the extreme voltage, which is not typical. The root mean square (rms) is a particular way of calculating a workable average. It would be easier to use the arithmetical average, but the average value of a sinusoid is zero, because every positive value is exactly balanced by a negative value. They all add up to zero. To find the rms, we first square each value (squares are all positive, so they can not add up to zero), then add them and finally calculate their average (mean square). The units of this quantity will be the square of the original units. For example, a mean square of voltages has the unit volts-squared. To get back to the original units, we take the square root of the mean (root mean square). For a sine wave, it can be shown that the rms value is 0.707 times ($1/\sqrt{2}$) the amplitude.

The rms value is of interest because it tells us the steady (DC) voltage that dissipates in a resistor the same power as the sinusoid. For example, a sinusoid of amplitude 1 V or a DC level of 0.707 V both dissipate the same power.

Noise characteristics vary with the physical nature of the component. Resistors generate only Johnson noise, though certain types of resistor which are composed of semiconductor material generate flicker noise too. There is no shot noise in a resistor. Capacitors have no significant resistance and inductors have very low resistance so neither of these devices contribute significantly to the noise levels in a circuit.

Semiconductors make a considerable contribution to noise levels. There are two sources of noise in an FET. Since the channel has resistance we find Johnson noise generated there. The higher the drain current, the greater the noise. There is also flicker noise, of importance at low frequencies, for example in audio amplifiers when they are operating at the lower end of the audio spectrum. There is very little shot noise in a JFET because the only junction is the gate and the current passing through this is the extremely small leakage current. There are no junctions in MOSFETs so these have no shot noise. BJTs and diodes have Johnson noise from the semiconductor material, shot noise from the pn junctions and flicker noise.

Low noise amplifiers

Noise is added to the signal at all stages of amplification but the most significant is that added at the early stages, for this is amplified most and accounts for the majority of noise present at the output. The output includes a component of noise generated even before the signal enters the amplifier. This comes from the signal source itself, whether it is a microphone, a sensor of some other kind, or a pre-amplifying stage. The source has resistance, apparent to the amplifier as output resistance, and this is a source of Johnson noise, which is voltage noise. To this we must add noise generated in the input stage of the amplifier. This may consist of both voltage and current noise. It can be shown that, for minimal noise from the source, there is an optimal value of source resistance. This is equal to the square root of the rms noise voltage divided by the rms noise current.

An obvious step to achieve low noise is to employ a low-noise transistor, particularly in the first stage. These are BJTs such as the BC109 specially manufactured to have low flicker noise. Transistors with high gain (such as the BC109) are also an advantage since they need a smaller base current. This means less shot noise and less flicker noise in the base-emitter junction. Often we use an FET in the first amplifying stage, as in the amplifier in Fig. 6.18. FETs have virtually no gate current (compared with a BJT which has appreciable base current through the biasing resistors) so no Johnson noise is produced. JFETs also have less shot noise and flicker noise as there is only a small leakage current at the gate. MOSFETs have no shot noise as there is no pn junction to cause it. Many of them have low channel resistance and therefore low Johnson noise. But they may have relatively high flicker noise at low frequencies.

If the first amplifying stage is a BJT, the voltage and current noise depend on the size of the quiescent collector current. The greater the current the greater the noise of both kinds. But, as collector current increases, current noise increases more than voltage noise. The current noise operates across the source resistance so the increasing effect of current noise at high collector current may be compensated for by reducing the source resistance. This intricate relationship between source resistance and collector current must be considered when designing low-noise amplifiers. For lowest noise, the collector current should be rated in tens of microamps, in which case the source output resistance should be tens of kilohms. If a higher collector current is essential, then the source resistance must be decreased.

There are other ways in which noise may be reduced. The effect of noise from the source is limited if the amplifier has a restricted bandwidth. Noise exists throughout the bandwidth of the source but, if only a part of this bandwidth is to be amplified, there is no point in admitting and amplifying noise of lower and higher frequencies, for all of this contributes to the degradation of the output signal. In other words, the bandwidth of the amplifier is best restricted to those frequencies that we really want to amplify. An audio amplifier should therefore have a bandwidth ranging from about 20 Hz to 20 kHz. Radio frequency amplifiers similarly need only a restricted bandwidth, just enough to include the carrier frequency and the sidebands.

Keeping up?

5. Name some sources of interference that might appear at the output of an amplifier.

6. Name the main sources of noise in electronic circuits and state their characteristics.

7. What are the main types of noise associated with BJTs, JFETs and MOSFETs.

A practical example

We will consider the multistage amplifier illustrated in Fig. 9.3. Before discussing its noise aspects, it is interesting to compare this amplifier with the similar amplifier in Fig. 6.16. Both comprise two BJT common-emitter stages. Both have negative feedback from the output to the emitter of Q1 and from the emitter of Q2 to the base of Q1. Fig. 6.16 has direct coupling of the stages, but they are capacitatively coupled (C2, C3, C4) in Fig. 9.3. Which type of coupling is used depends on quiescent voltage levels at the collector of Q1 and the base of Q2. Capacitor coupling may also be used to shape the frequency response of an amplifier.

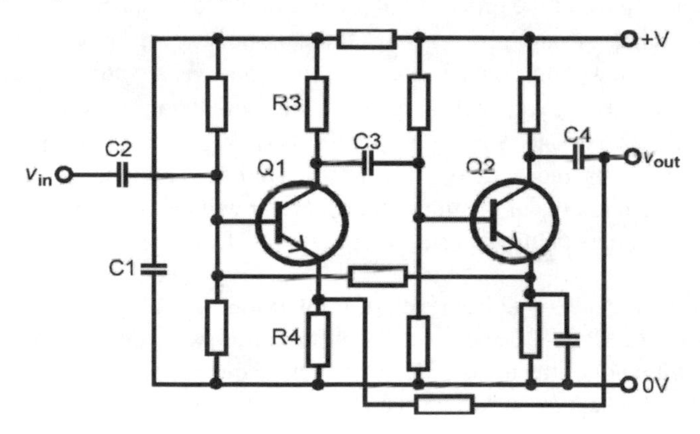

Figure 9.3 *A multistage BJT amplifier (compare with Fig. 6.16) is the subject of attempts at noise level reduction.*

The amplifier in Fig. 9.3 differs from that in Fig. 6.16 in having a resistor in the supply rail and a capacitor (C1) between the positive rail and the 0 V line. These act as a low-pass filter to reduce the feedback of signals from the output stage to the first stage of amplification (Q1). This is not essential in the amplifier as shown in the figure, but the two stages of this amplifier were taken from a 3-stage amplifier with a Class B power output stage (similar to Fig. 7.4). The powerful output stage draws large and varying currents from the power-lines, causing fluctuations in the voltage. These would affect the operation of the earlier stages were it not for the resistor and capacitor acting as a lowpass filter to remove them. This is an example of low-noise (or perhaps we should call it low-interference) design.

In Fig. 9.4 the amplifier is tested on a simulator to calculate the noise present at every frequency in the range 10 Hz to 1 MHz. All sources of noise within the circuit (including the signal source) are examined for noise by the simulator and their contributions to the output noise are summed.

Resistors produce Johnson noise, transistors produce all three kinds of noise, but capacitors and inductors are noiseless. Since all values are rms values, we can not simply add them, arithmetically. Instead, we square them, *then* add them and finally take the square root of the total. This gives the noise density in volts per root hertz. As expected, the noise density is constant over almost the whole range of measurement. Its value in this circuit is 5.4 μV/√Hz (we will leave out the √Hz in future). This means that peaks of voltage of the order of 5.4 μV are likely to be superimposed on the output signal. At any given instant there is no way of knowing what the noise voltage will be, since the generation of noise is a random process, but 5.4 μV is a likely average. It may be a *little* more than this or a *little* less, it is not likely to be *much* more or *much* less, though it is not impossible for large deviations to occur from time to time. Whatever happens, if the signal level at the output is in the microvolt range, it will be severely degraded by the noise. It may be impossible to separate one from the other.

Another measure of noise level is total noise, useful when we want to compare the noise in one circuit configuration with that in another. To obtain total noise, the noise density at each frequency is integrated from the lowest to highest frequency in the plotted range. For 10 Hz to 1 MHz the total noise of this amplifier is 2.90 mV. It would be extremely unlikely to happen, but a peak of 2.90 mV (but no greater) could appear on the output at any moment.

Now we will see what can be done to reduce the noise level. In this investigation we are shunning the use of mathematical equations and proceeding by trial and error (sometimes), using the simulator to evaluate the success or otherwise of our attempts. Since there seems to be a generally high level of noise across the whole spectrum we might surmise that this is Johnson noise, which may well originate in the high output resistance of the signal source. For the test run used to produce Fig. 9.4, the signal source is set to have an output resistance of 100 kΩ. If we reduce the output resistance from 100 kΩ to 1 kΩ, and repeat the analysis, we obtain Fig. 9.5. There is a distinct fall in noise levels across most of the range, though we are left with a peak at low frequencies. Possibly this is flicker noise, which dominates the noise at low frequencies, especially when other sources of noise have been made less important. The maximum noise density has now been reduced to 2.78 μV and the noise is only 1.32 μV over a large part of the range. The total noise at 1 MHz is only 980 μV, slightly more than one third of the original total. From this we conclude that

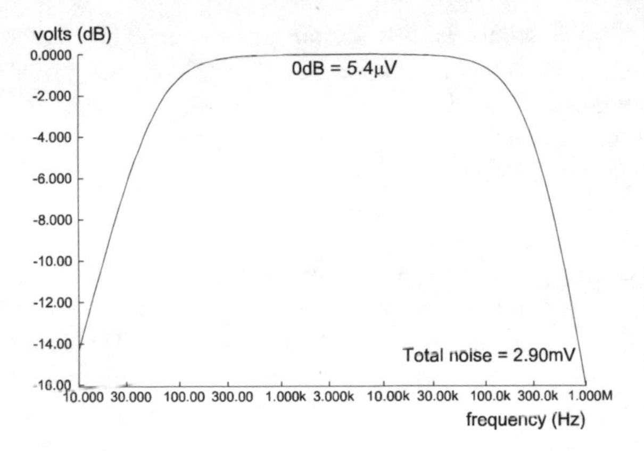

Figure 9.4 *A plot of noise level against frequency shows that the amplifier has significant noise levels over a range from 10 Hz to 1 MHz.*

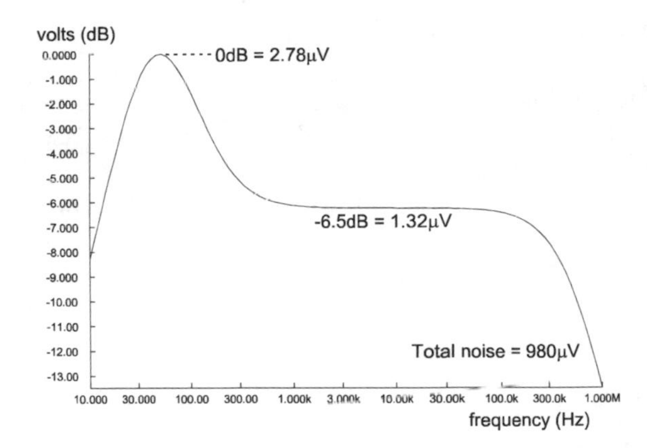

Figure 9.5 *Reducing the source output resistance gives major noise reduction, particularly above about 300Hz.*

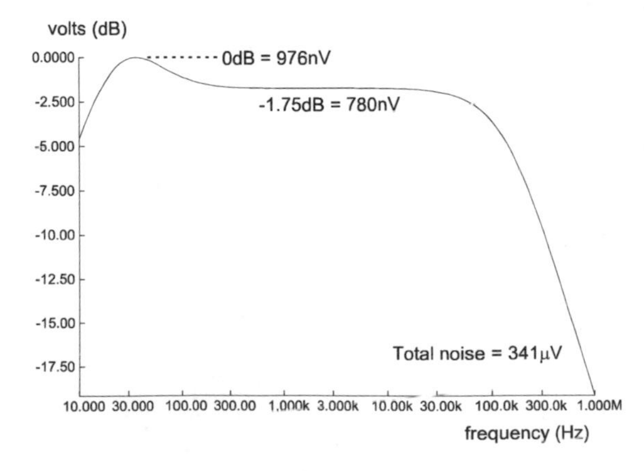

Figure 9.6 *If the collector current of Q1 is reduced there is a further major noise reduction, particularly below 100 Hz.*

the source should have a relatively low output resistance. This rather dramatic alteration of source output resistance from 100 kΩ to 1 kΩ would normally be the preliminary to adjusting it more carefully to find the optimum value. But for this demonstration we will leave it as it now is, and carry on with other steps toward noise reduction.

In earlier discussion of noise we stated that noise is related to collector current. Q1 is most concerned here since any noise this produces is greatly amplified before it reaches the output. When we measure the quiescent collector current of Q1 on the simulator we find that it is 74 μA. This is not large as collector currents go, but large enough for the first stage in an amplifier. A simple way to reduce collector current is to increase the values of R3 and R4 while keeping them in the same ratio. Increasing the values results in a reduction of the current to 10 μA. Keeping source resistance at 1 kΩ, the noise analysis produces the results shown in Fig. 9.6.

The peak of noise at low frequencies is now almost gone. We are now measuring noise density in nanovolts. Note too that not only is the maximum noise level less than before, but the plot goes down to –17.5 dB instead of –13 dB as in Fig. 9.5. The plot indicates a much greater falling-off to lower noise levels. The total noise is reduced to 341 μV, which is a very satisfactory figure for many applications. Once again, we might adjust the collector current, and perhaps readjust the source resistance to find optimum values.

If the amplifier is to be used over a limited frequency range there is no point in having a wide bandwidth. The wider bandwidth just lets more noise through to the output. Suppose that we intend the bandwidth to extend from 300 Hz to 30 kHz. Cutting off the lower end of the bandwidth is important, since this often contains large amounts of flicker noise. In this example, the amplifier already has reasonably good cut-off at this end. At the other end, if the bandwidth is not to extended as far as 1 MHz, we can remove higher-frequency of noise with a low-pass filter. The simplest possible filter is a passive RC filter as shown in Fig. 9.7. The values are calculated for a cut-off at 30 kHz. The figure shows a two-pole filter and in Fig. 9.8 we plot the effect of feeding the amplifier output through this filter. We have plotted noise levels at the amplifier output, at the output of the first filter pole and at the output of the second pole. The graph shows the expected fall-off of noise above about 30 kHz. Using a single filter pole reduces the total noise to 33 μV. Although the second filter makes an apparently big

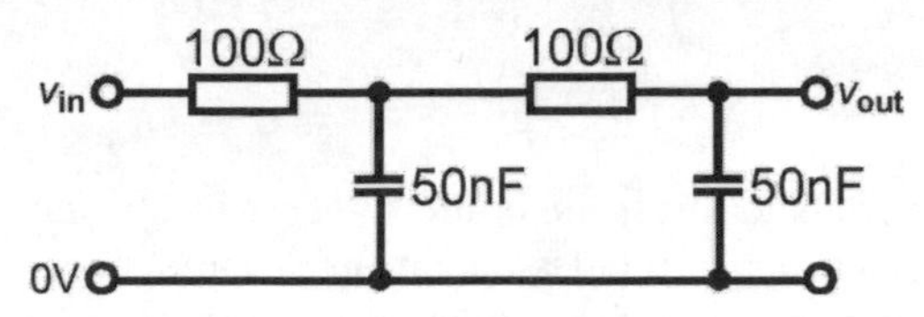

Figure 9.7 *A 2-pole passive lowpass filter, f_C = 30 kHz.*

difference in Fig. 9.8, this is merely a proportional reduction on a level that is already low (remember decibels are ratios). The total noise remains close to 33 μV so a second stage of filtering is not worth installing. It is apparent that reduction in bandwidth results in significant reduction of noise. Reduction in bandwidth by simple filtering such as we have just used is limited by the fact that roll-off does not eliminate the out-of-band frequencies entirely. It might be better to narrow the bandwidth to reduce noise further, even if this means introducing distortion into the signal.

This covers all that we would normally do to reduce noise. The overall result of the recommended changes has been to reduce total noise from 2.9 mV to 33 μV, almost a ninety-fold reduction. There still remains another approach to the problem. Noise is temperature dependent, so keeping the amplifier cool helps reduce noise. Taking the opposite viewpoint, we should prevent it from getting hot, either by using adequate heat-sinks or, more positively, by using cooling fans. Instrumentation amplifiers need cooling if they are to achieve maximum sensitivity to small signals. As an illustration of this, we cool our simulated circuit from its standard operating temperature of 27 °C down to freezing point, 0 °C. The result is to reduce total noise from 33 μV to 23 μV.

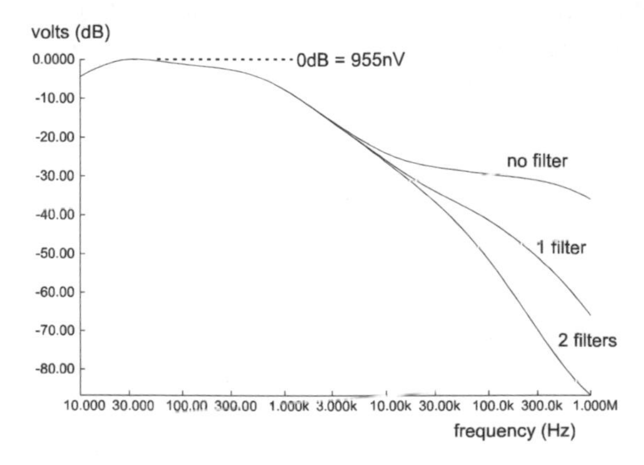

Figure 9.8 *Lowpass filtering of the amplifier output removes much of the high-frequency noise.*

Summing up

Non-linearity in an amplifier leads to distortion of the signal, which results in the production of harmonics. This includes distortion of large signals due to clipping. Harmonics give rise to intermodulation products. Even if these are outside the intended operating range of the amplifier they represent a waste of power.

Group delay is a measure of the rate of change of phase delay with frequency at a given frequency. Zero or low group delay is essential for least distortion.

Noise is an unwanted signal superimposed on the amplified signal. It may consist of interference from external sources and random noise signals due to the nature of electrical conduction.

Noise comprises:

- Johnson noise: random voltage changes due to thermal motion of charge carriers. Generated in any resistance and is proportional to temperature and current.
- Shot noise: random current changes due to irregularities in the flow of charge carriers. Generated in semiconductors. Shot noise is more important when currents are small.
- Flicker noise: random current changes of unknown cause. Generated in semiconductors and in thermionic valves. Amplitude is inversely proportional to frequency so this is also known as *1/f noise.*

Measures to reduce noise include:

- Design for resistors of small value.
- Design for small collector and drain currents.
- Use special low-noise semiconductors.
- Adjust the output resistance of the signal source to a suitable value.
- Keep the circuit, especially the semiconductors, as cool as possible.
- Pay special attention to minimising noise in the first stage of amplification.
- Use an FET in the first amplifying stage.
- Limit the bandwidth of the amplifier to cover only the essential range of frequencies.

Test yourself

1. Explain the causes of harmonic distortion and ways in which its effects can be minimised.

2. Having designed an amplifier, what points would you check on if you wanted to make the amplifier as noise-free as possible?

10 Operational amplifiers

The purpose of an operational amplifier is to perform a mathematical *operation*, such as addition, subtraction, multiplication or division. An operational amplifier (or *op amp*, as we shall call it from now on) can be built from discrete components but nowadays comes in the form of an integrated circuit. Many types of op amp exist with various specifications to suit them to different applications but the principles of operation are common to all types. There are also many other kinds of ic amplifiers, specialised for use as audio amplifiers, instrumentation amplifiers, servo amplifiers and other purposes, but these designs are too varied, too numerous, and of too limited general interest to be covered in this book.

Nowadays we tend to think of mathematical operations as being the prerogative of digital circuitry. Op amps are strictly analog operators. They were first developed for use in building analogue computers, in the days before the digital form took over the field. But there are many instances when a precise fast-acting circuit is needed to perform just a few steps in a 'calculation' (such as detecting and multiplying the output from a light sensor, or turning an AC signal into its precisely rectified equivalent).

Op amps and others

In the discussions in this chapter the term 'op amp' refers to an integrated circuit, represented by the symbol in Fig. 10.1. We do not refer to this circuit unit as an 'amplifier', even though it can amplify when suitably connected. Many of the circuits in the chapter consist of an op amp connected to additional components such as resistors and capacitors to make up a circuit with a specific function. This is what we mean when we refer in this chapter to an 'amplifier'.

This role is ideally filled by the op amp. There are so many uses for op amps that most types are manufactured in extremely large quantities and are consequently very cheap. They are so easy to use that we tend to employ them for relatively trivial tasks, far removed from their original role of precision calculators.

Op amps belong to three main families, depending on the technology used in their construction. The original op amps were based on BJTs, and this technology still provides a wide range of op amps for widespread application. There are more types in this family than in the other two. Op amps with JFET inputs and CMOS inputs are also common. Each family has its advantages and disadvantages, as is explained below. There is a fourth family, the 'bifet' op amps, which combine most of the advantages of all families by having a FET input stage, followed by BJT amplifying stages.

Op amp features

An op amp typically has two inputs, a single output and two power-line connections (Fig. 10.1). Most op amps work on a dual supply, such as ± 9 V or ±15 V, and some may operate on voltages as small as ±1 V. Some types can be used on a single supply. This is because they are able to accept input levels as low as that of the negative rail. In general, input levels must be between those of the two power rails and, with many op amps, within a few volts inside these limits.

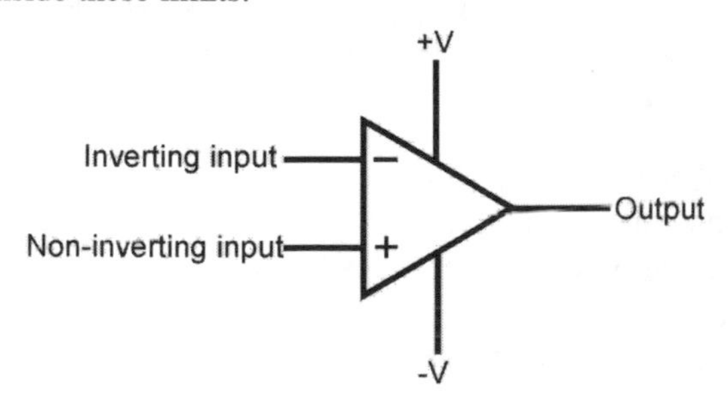

Figure 10.1 *A basic op amp, seen as a black box, has two inputs and one output.*

Internally the inputs are connected to a differential amplifier (see Chapter 6) so they work in opposite senses, one being the *inverting input* (indicated by '–' on the symbol, the other being the *non-inverting input* (indicated by '+'). An increase of voltage at the inverting amplitude causes the output voltage to fall. An increase of voltage at the non-inverting amplitude causes the output voltage to rise. If both inputs are at the same voltage, the output is 0 V (but see later). From now on we simplify the descriptions by calling the inverting input the (–) input and the non-inverting input the (+) input.

Because the op amp is essentially a differential amplifier, the most important fact about input signals is not whether they are positive or negative relative to the 0 V rail but whether they are positive or negative of each other. It is the voltage *differential* that determines the output. So, if the (–) input is negative of the (+) input, the output is positive. If the (–) input is positive of the (+) input, the output is negative. Output voltage swing may be limited. The output voltage usually does not swing as far as the supply rails. Typically, it swings between about 2 V below the +V levels and 2 V above the –V level. If a full swing from + V to – V is required, most CMOS op amps can do this.

The size of the output voltage, whether positive or negative, is proportional to the difference between the inputs by an amount known as the *open loop gain.* The open-loop gain of a typical op amp is in the order of 106 dB, that is, a voltage gain of 200 000. A gain as large as this is advantageous if the op amp is being used as a comparator, to tell which of its two inputs is at the higher voltage (see later) but it is not much practical use for measuring voltage differentials. With a typical maximum output swing of ±13 V, any input difference greater than 65 µV is off the scale. We use negative feedback to reduce the gain of the amplifier to manageable and (more important) to precisely known levels, as explained later.

Another notable feature of op amps is their input resistance, which is high. Even those which have a BJT input stage have typical input resistance of 2 MΩ, while those based on JFET and MOSFET technology have input resistances rated at 1 teraohm (1 TΩ = 10^{12} Ω). By contrast, op amps have a low output resistance, usually about 75 Ω. Most op amps have a robust construction and, in particular, are able to withstand short-circuiting of their output for an indefinitely long period.

There are certain other features, some common and some restricted to certain types, that must be thought about when selecting an op amp for a particular purpose.

Other features

Real op amps fall short of the ideal in a number of ways:

- *Input bias current* The amplifier requires a minimum current at its input before it will begin to operate. In BJT op amps, this is a few nanoamps, and in FET op amps it is a few picoamps, which represent leakage current at the input gates. Normally we can ignore these currents when designing a circuit.

- *Input offset current* Even though the op amp is built on a single chip there are differences between the integrated components at the two inputs. Therefore, the inputs draw different amounts of current even if they are supplied from sources of equal output resistance. There may be a difference of as much as 50 nA in BJT op amps, but much less in FET op amps.

- *Input offset voltage* In the ideal op amp the output is 0 V when the inputs are at exactly equal voltages. In the practical op amp this is not necessarily so. Equal voltages produce an output slightly greater or less than 0 V. The voltage difference between inputs needed to produce an exactly zero output is known as the input offset voltage. This may be about 10 μV in precision op amps or as much as 1 mV in other op amps. In general amplifiers with BJT input have the lowest input offset voltage so are best suited for precision applications. If the voltages in our amplifier are rated in tens or hundreds of millivolts, we may

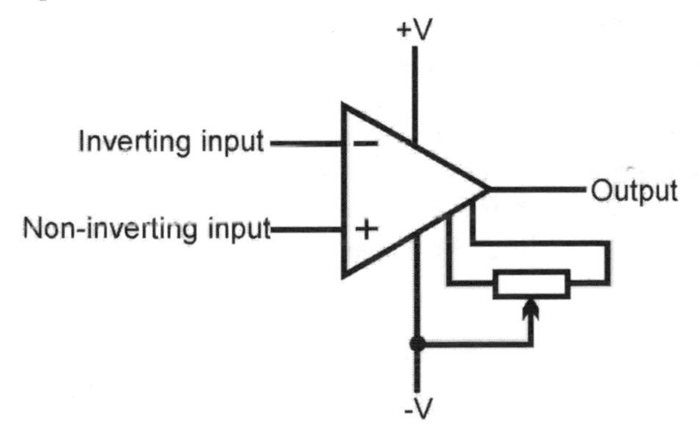

Figure 10.2 *Input offset voltage errors are usually compensated for by using the offset null terminals.*

usually ignore input offset voltage, and we can ignore it in certain other circumstances too. In precision circuits we may need to take it into account by *nulling* it, that is by adjustments which produce a 0 V output for 0 V input differential. This is usually done by connecting a variable resistor to two terminals on the integrated circuit, known as the *offset null* terminals (Fig. 10.2). Some ics do not have these terminals, especially if there are 2 or 4 op amps on the same chip.

- *Common mode gain* Discussion of differential amplifiers in Chapter 6 indicates that ideally this is zero. In practice, it is around 2 for many op amps. This is very small in comparison with the open loop gain of the op amp, so that the *common mode rejection ratio* (CMRR) is typically as high as 80 dB to 100 dB. CMOS op amps have particularly high CMRRs.

- *Slew rate* describes the ability of the op amp to keep up with changes in input voltages. It is expressed as the maximum rate at which the output voltage can swing, in volts per microsecond. Most op amps have slew rates in the region of 1 V/μs to 10 V/μs, though high-frequency op amps have slew rates as high as 600 V/μs. With low slew rate and a high frequency sinusoid, the output is forever trying to keep up with the signal. One effect is that output amplitude is reduced and another is that the signal tends to become a triangular wave, introducing severe distortion. CMOS op amps in general have low slew rates so are unsuited to high-speed operation.

- *Gain-bandwidth product.* As in all amplifiers, gain falls off with increasing frequency. A typical op amp has its full open-loop gain from DC up to about 10 kHz (the *full-power bandwidth*), but gain begins to fall above that frequency. Often it falls to unity at a frequency of 1 MHz and this is said to be the gain-bandwidth product or transition frequency of the op amp. Not only does gain fall off but also the phase of the output signal lags further and further behind that of the input. Lag approaches –180° as gain approaches unity.

Op amp circuits include resistors and semiconductors so it is inevitable that they are sources of noise. The noise figure for BJT op amps is around 200 nV/√Hz, 15 nV for FET-based op amps and 50 nV for MOSFETs.

The next step is to look at a number of conventional circuits based on an op amp.

Keeping up?

1. List the terminal connections of an op amp.
2. Quote typical values for the input and output impedances of an op amp.
3. Explain what is meant by (a) input offset current, (b) input offset voltage, and (c) slew rate.
4. Quote typical values for (a) the open loop gain, and (b) the common mode gain, of an op amp.

Inverting amplifier

The inverting amplifier (Fig. 10.3) makes use of negative feedback. To begin with we may assume that since the inputs take virtually no current, there is no current through R_B and its ends are at the same voltage, 0 V. A positive voltage v_{IN} is applied to the input, causing a current to flow along R_A toward the (–) input. This input too admits no current so the *same* current must flow on through the feedback resistor R_{FB} and eventually enter the output terminal of the op amp. Let us examine the effect of the voltage at the (–) input. If it is positive of the (+) input, the output voltage falls, pulling down the voltage at the (–) input. If it is negative of the (+) input,

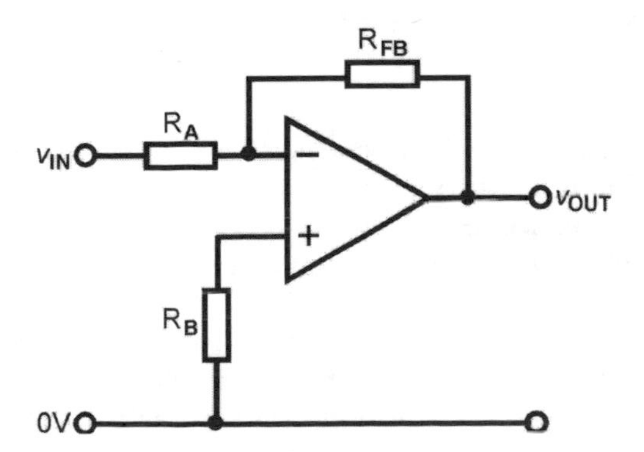

Figure 10.3 *An inverting op amp amplifier has its output joined to its negative input. For simplicity, the power connections are omitted in this and later figures in this chapter.*

the output voltage rises, pulling up the voltage at the (–) input. The feedback acts to as to raise or lower the voltage at the (–) input until it is *equal* to the voltage at the (+) input. Because the (+) input is at 0 V, the (–) input, too, is brought to 0 V. The circuit is then stable.

R_A has a voltage v_{IN} at one end and 0 V at the other (Fig. 10.4). So the current through it is $i = v_{IN}/R_A$. The same current flows through R_{FB}. From this we deduce that the voltage drop along R_{FB} is $R_{FB} \times i = R_{FB} \times v_{IN}/R_A$. Since one end of R_{FB} is at 0 V, its other (output) end must be at $-R_{FB} \times v_{IN}/R_A$. This is v_{OUT}, the output voltage of the circuit. The gain of the circuit is:

$$\frac{v_{OUT}}{v_{IN}} = -\frac{R_{FB}}{R_A}$$

For example, if R_{FB} = 22 kΩ and R_A = 1.2 kΩ, the gain is –22/1.2 = –18.3.

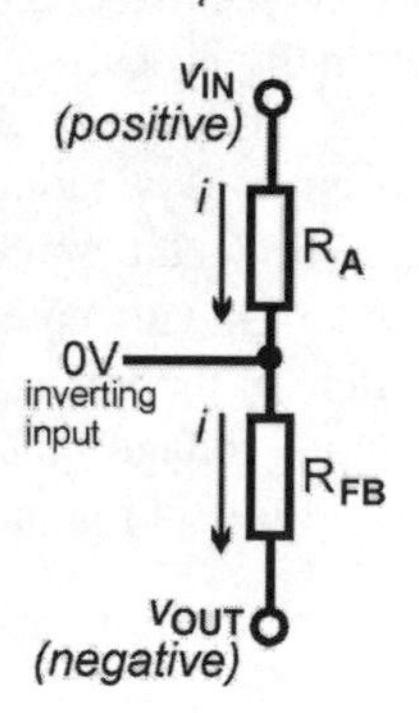

Figure 10.4 *In an inverting amplifier, if v_{IN} is positive, current flows through R_A toward the inverting input but does not enter it. The same current then flows through R_{FB} to the negative voltage at the output terminal of the op amp. The current is the same through each resistor, so the voltage drops across the resistors are proportional to their resistances.*

Note that the gain of the circuit depends only on the values of the two resistors and can be set very precisely by using high-precision resistors. Note too that it is far less than the open-loop gain of the op amp itself. It is known as the *closed-loop gain* of the amplifier.

Resistor R_B is necessary in the circuit to compensate for input offset current. When we said that no current flows into the (+) input, this ignored the fact that there really is a small current entering the inputs of BJT op amps. The current flows through R_B, bringing the voltage at the (+) input slightly below 0 V. It produces an *offset voltage*. Similarly, there is current flowing into the (–) input. Consequently, not *all* of the current through R_A actually flows on through R_{FB}. It is as if some of the current flows into the input through R_A and some flows back and into the input through R_{FB}. In this respect R_A and R_{FB} are in parallel and we compensate for input offset current by making R_B equal to R_A in parallel with R_{FB}. Then both inputs are subject to the same offset voltage. For example, if $R_A = 1.2\ k\Omega$ and $R_{FB} = 22\ k\Omega$ as above, then $R_B = 1.14\ k\Omega$. For practical purposes it is often sufficient to make R_B equal to R_A, especially when R_{FB} is large, as it usually is. If the op amp has FET inputs, input offset current may be ignored and the (+) input is wired directly to the 0 V line.

Virtual earth

When the op amp is connected as an inverting amplifier, the circuit operates to bring the (–) input to 0 V (ignoring the offset voltage). Whatever the value of v_{in}, the op amp adjusts its output (within the limits in which is can operate, of course) to bring the (–) input to 0 V. The circuit *behaves* as if the (–) input is connected directly to the 0 V line, or earth. We say that the (–) input is a virtual earth. This property is important in some of the circuits that we examine later.

When an op amp is wired as a medium-gain or high-gain amplifier, a small value is chosen for the input resistor R_A. For example, to obtain a gain of 200, we might make R_B equal to 2 MΩ and R_A equal to 10 kΩ. Since the input terminal *of the amplifier* is connected through R_A to a virtual earth, the input resistance of the amplifier equals R_A. In the example, the input resistance *of the amplifier* is only 10 kΩ, even though the input resistance *of the op amp* is measurable in megohms or even teraohms. Low input resistance tends to be a drawback of inverting amplifiers in certain applications.

Non-inverting amplifier

The prime advantage of this amplifier (Fig. 10.5) is that the signal is sent directly to the (+) input and to nowhere else. The input resistance is therefore equal to the full input resistance of the op amp. As in the previous amplifier, R_B is there to compensate for voltage offset, but it is not necessary with CMOS and FET input op amps and can be omitted. If

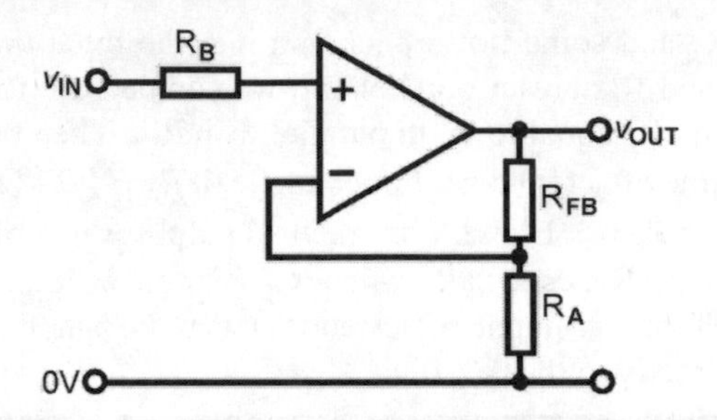

Figure 10.5 *A non-inverting op amp amplifier has its output connected to a potential-divider. This taps part of the output, which is then fed back to the inverting input.*

included, it is made equal to R_A and R_{FB} in parallel.

The output of the op amp goes to a potential divider consisting of R_{FB} and R_A, grounded at the other end. The divider is tapped between the two resistors so that the voltage fed back to the (–) input is $v_{OUT} \times R_A /(R_A + R_{FB})$. Ignoring voltage offset, the op amp is stable when the voltages at its two inputs are equal. Then $v_{IN} = v_{OUT} \times R_A /(R_A + R_{FB})$. The gain of the amplifier is positive, and given by:

$$\frac{v_{OUT}}{v_{IN}} = \frac{R_A + R_{FB}}{R_A}$$

A special version of this amplifier has R_A infinitely great and R_{FB} equal to zero (Fig. 10.6). In the limit, the gain equation then simplifies to $v_{OUT}/v_{IN} = 1$. It is a *unity gain voltage follower,* with a similar action to the emitter-follower and source-follower circuits. It has an input resistance of several megohms or teraohms, and an output resistance of (usually) 75 Ω. This makes it very useful as a buffer between a circuit with high output resistance (many kinds of sensor fall into this category) and one with low input resistance.

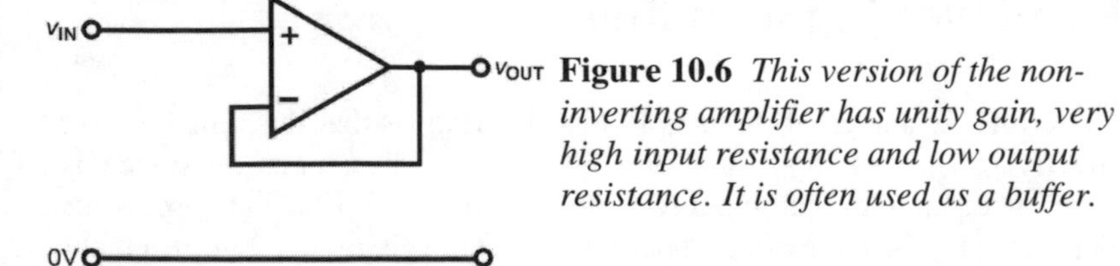

Figure 10.6 *This version of the non-inverting amplifier has unity gain, very high input resistance and low output resistance. It is often used as a buffer.*

Summing amplifier

This is also known as an adder or mixer amplifier and makes use of the virtual earth of the standard inverting amplifier. Since all input currents seem to disappear into the virtual earth (actually they flow on through the feedback resistor), we can supply several different currents to the amplifier simultaneously, through separate resistors (Fig. 10.7). Whatever current flows through each input resistor, the other end of the resistor remains at 0 V. The currents affect the amplifier independently.

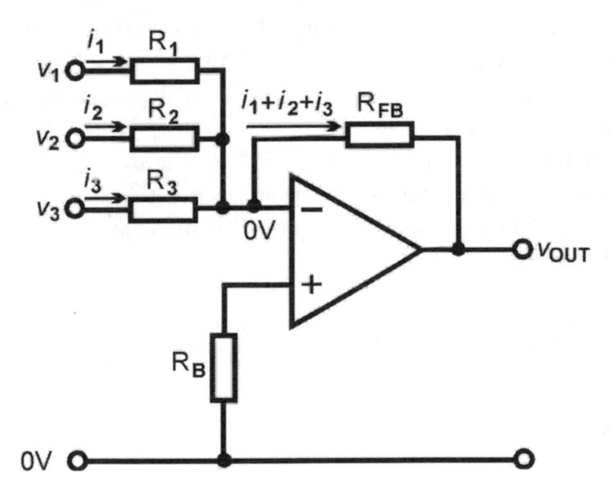

Figure 10.7 *If an inverting amplifier has several inputs, the currents flowing in through each are combined (added or mixed) as they flow on through the feedback resistor.*

There are a number of different ways in which this kind of circuit may be used. If all resistors including R_{FB} are equal (to R), then $i_1 = v_1/R$, $i_2 = v_2/R$, and $i_3 = v_3/R$. The list of equations may be extended if there are more than 3 inputs.

The current flowing through the feedback resistor is:

$$i_1 + i_2 + i_3 = (v_1 + v_2 + v_3)/R$$

The voltage drop across R_{FB} is:

$$-\,(total\ current) \times R = -(v_1 + v_2 + v_3)$$

In words, the output voltage is the negative of the sum of the input voltages. Although we have considered the case in which all input voltages are positive, it is allowable for one or more inputs to be negative.

Another frequently-used technique is to weight the resistors on a binary scale. For example, we could make $R_1 = R_{FB}$, $R_2 = R_1/2$ and $R_3 = R_2/2$. Voltage inputs are equal to v when switched on, but any or all can be switched off and are then equal to zero. Suppose for example, that $R_{FB} = 4$ kΩ, $R_1 = 4$ kΩ, $R_2 = 2$ kΩ and $R_3 = 1$ kΩ. Suppose also that the standard ON input voltage is 1 V. If 1 V is applied only to R1, $i_1 = 1/4000 = 250$ μA. This current passing through R_{FB} brings the output end to −250 μA × 4000 = −1 V. If 1 V is applied only to R_2, $i_2 = 1/2000 = 500$ μA. Output falls to −500 μA × 4000 = −2 V. Thus a unit input to R_1 represents the binary number 001 (=1 decimal). A unit input to R_2 represents 010 (= 2 decimal). If 1 V is applied simultaneously to R_1 *and* R , the individual currents are 250 μA and 500 μA as before, but now 75 μA flows through R_{FB}, and output falls to −750 μA × 4000 = −3. This is equivalent to a binary input of 011 (=3 decimal).

Using all three inputs, we can represent all binary numbers from 000 to 111, with outputs ranging in steps of 1 V from 0 V to −7 V. In other words, this amplifier with weighted inputs is a simple but workable digital-to-analog converter.

Comparator

An operational amplifier is used as a comparator by feeding two voltages directly to its two inputs (Fig. 10.8). As explained earlier in the general discussion of op amp behaviour, if v_1 is greater than v_2, the output swings low. If v_2 is greater than v_1, the output swings high. In theory, there is a low

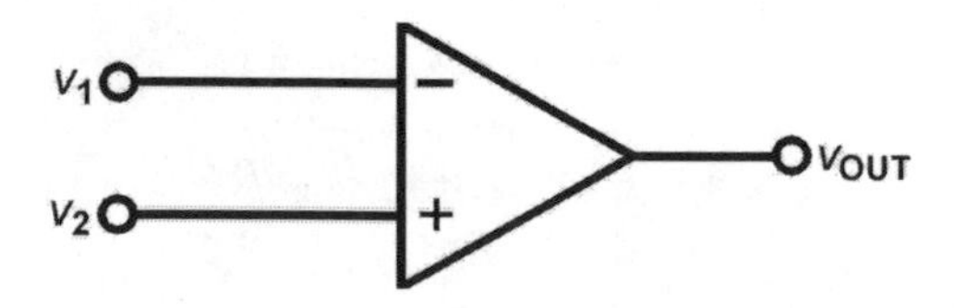

Fig 10.8 *An op amp may be used as a comparator to indicate which of two input voltages is the higher.*

or zero output if the inputs are close or equal, but in most applications the output swings as far as it will go toward one power rail or the other. It can often be used as the input to a digital logic circuit. It is important for the op amp to have low input offset voltage, otherwise it may give an incorrect indication for two close voltages. When used in a digital circuit it may be important to use an op amp with high slew rate.

Differential amplifier

Op amps are basically differential amplifiers and in Fig. 10.9 we see one being used for that purpose. Considering the two inputs separately, we first note that v_2 is being applied to a potential divider consisting of R_A and R_B, where $R_B = R_{FB}$. Input v_1 is fed to the (–) input and there is feedback so the op amp settles when the voltages at the two inputs are equal. In this circuit the (+) input is *not* at 0 V, so there is no virtual earth at the (–) input. By applying the same techniques that we have used with the other circuits, it can be shown that:

$$v_{OUT} = \frac{(v_2 - v_1) \times R_{FB}}{R_A}$$

The output is proportional to the difference between the input voltages.

A serious flaw in this circuit is that the inputs have different input resistances. Signal v_1 sees an input resistance R_A, for the same reason as in

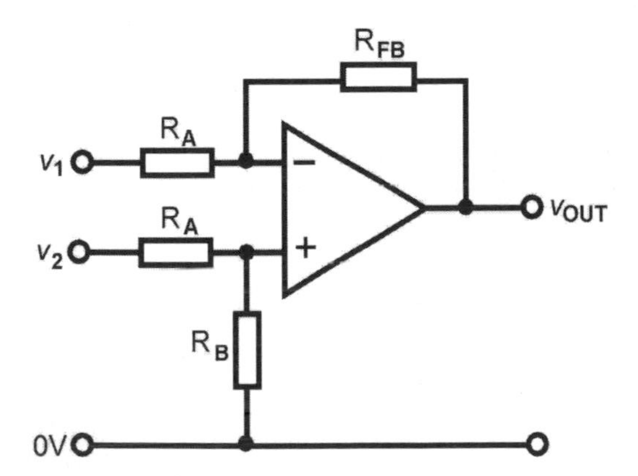

Figure 10.9 *The output of this differential amplifier is proportional to* $(v_2 - v_1)$.

the inverting amplifier. Signal v_2 sees an input resistance of R_A in series with R_B, so the combined input resistance is finally more than R_A. With unequal input resistances, power transfer from sources with equal output resistance will be unequal and the amplifier will not measure the differential voltages precisely. On top of this, if R_A is given a low value to increase gain, both inputs have low input resistances. This will introduce errors if the source has high output resistance. A more precise differential amplifier is described in Chapter 11.

A useful variation of this circuit is to give all the resistors equal value. Then the equation simplifies to $v_{OUT} = v_2 - v_1$. In this form the circuit is also known as a *subtractor*.

Differential amplifiers are often used to compare low voltages such as those produced by a strain gauge or by thermocouples. These may often share common-mode interference signals which are eliminated by using a differential amplifier.

Differentiator

Although this amplifier (Fig 10.10) has a name similar to that of the amplifier described in the previous section, their actions are entirely unalike. The differentiator performs the mathematical operation of *differentiation*. Differentiation is one of the operations of the calculus, and its purpose is to calculate *rates of change*. Given a voltage that is changing at

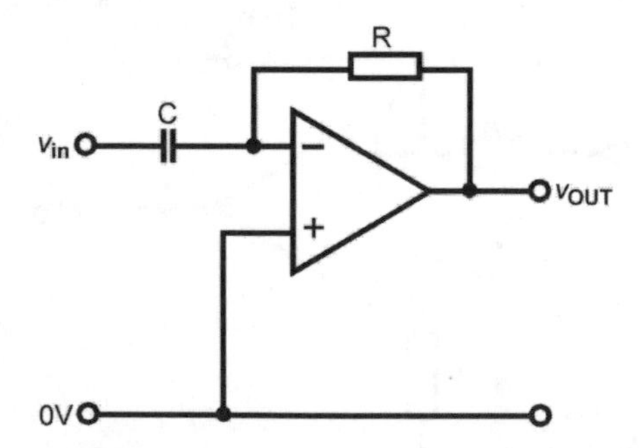

Figure 10.10 *The circuit of a differentiator is similar to that of an inverting amplifier,but the capacitor makes the op amp responsive to voltage changes instead of to voltage levels.*

Rates of change

If a quantity such as a voltage v changes in time, we express the rate of change at any instant by using the symbol dv/dt. Often the units of dv/dt will be volts per second. The 'd' in the symbol means 'an infinitely small change of', so the symbol means 'the infinitely small change of v that occurs during an infinitely small time t'. In other words, it is the instantaneous rate of change of v.

If we are plotting a graph of voltage against time, dv/dt is the *gradient* of the curve at any point.

a particular rate, which may itself be changing, the circuit calculates the rate of change of voltage at any instant.

In the differentiator, dv/dt represents the changing voltage across the capacitor. Given this rate of change, the rate of change of charge on the capacitor is $C \times dv/dt$, measured in coulombs per second. One coulomb per second is, by definition, a current of 1 amp, so the current flowing into or out of the capacitor is $i = C \times dv/dt$ amps. This current also flows through resistor R, causing a voltage to develop across it. One end of the resistor is at 0 V because the (–) input is a virtual earth. Therefore the voltage at the other end, the output voltage, is $v_{OUT} = -iR = -RC \times dv/dt$. If we measure v_{OUT} and know the values of R and C, we can calculate dv/dt.

Fig. 10.11 shows the input and output of an op amp differentiator circuit when it is processing a 100 mV, 1 kHz square-wave signal. Here it is plotted on a 5-times voltage scale. For much of its cycle, a square wave has constant value, either +100 mV or –100 mV. At these times the signal does not change, so its rate of change is zero and the output of the amplifier is 0 V. Twice during the cycle the input swings very rapidly from +100 mV to –100 mV, or the other way about. During that time its rise may be small, perhaps only 200 mV as in this example, but the time it takes to rise is only a few nanoseconds. The result is a high *rate of change* and at these instants the output of the amplifier shows tall positive or negative spikes. As indicated by the negative sign in the expressions in the previous paragraph, this is an inverting amplifier, so the spikes are positive when input falls and negative when input rises.

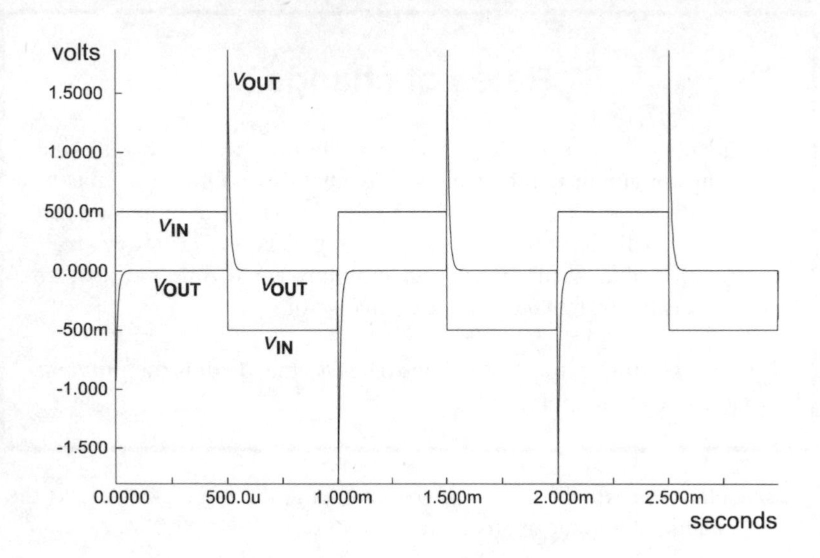

Figure 10.11 *When a square-wave signal is fed to a differentiator, it emerges as a series of sharp spikes coinciding with the times when the square wave is rising or falling.*

A differentiator is sensitive to any abrupt changes in the input signal, including those produced by noise. These are seen as numerous spikes on the output signal. This may be a problem at high frequencies because high-frequency peaks have the most rapid rates of voltage change. Even if noise levels are low, as measured in volts, the rate of change voltage is high at the high-frequency end of the noise spectrum, and large peaks are generated. The frequency response of the differentiator may be damped at high frequencies by including a resistor R_1 in series with the input capacitor C, and a capacitor C_1 in parallel with the feedback resistor R. Values are chosen so than $RC = R_1C_1$. Then the maximum gain of the circuit is R/R_1 at a frequency of $1/2\pi R_1C_1$.

A differentiator is used in measuring circuits where the rate of change of a quantity is of more interest than its absolute value. For example, we can determine the *velocity* of an object by measuring its *position*. Position might be measured directly by attaching the object to the wiper of a linear variable resistor and measuring the voltage between the wiper and one end of the resistor. If that voltage is fed to a voltmeter or oscilloscope, the reading indicates the position of the object along its track. If the voltage is first fed to a differentiator the output of this gives d*v*/d*t*, the rate of change of voltage with time. This is a measure of its rate of change of position with time, in other words, its velocity. At every instant the differentiator converts position to velocity. We can take this further by passing the velocity signal to a second differentiator. The output of this is proportional to the rate of change of *position* with time, which is *acceleration*.

Integrator

Integration is mathematically the inverse of differentiation. The circuit of an integrator (Fig. 10.12) shows that we have obtained the inverse action by swapping the resistor and capacitor of the differentiator. Given a positive input, current flows through R and charge accumulates on plate A of the capacitor. Any tendency of the voltage at plate A to increase, which would also raise the voltage at the (–) input, is countered by the action of the op amp, trying to make the two inputs equal. As the charge on plate A increases, the output of the op amp drops, pulling plate B negative, and this

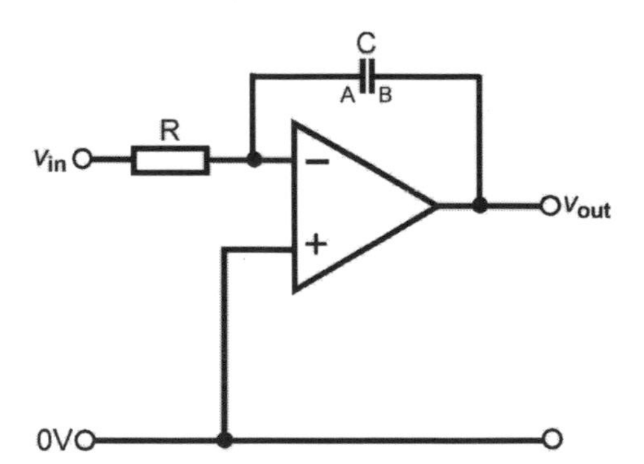

Figure 10.12 *In an integrator with positive input, current flowing through R charges the capacitor and the output falls.*

keeps the voltage on plate A equal to zero. For as long as the input voltage is constant and positive, a constant current *i* flows through R, and a constant current –*i* flows from plate B to the output of the op amp. The output voltage of the op amp falls steadily. The bigger the input voltage, the bigger the current and the faster it falls. The longer the time, the further it falls. So the amount of falls depends both on v_{IN} and the time for which it is applied. Fig. 10.13 shows the output when the input v_{IN} is a square wave. The curve

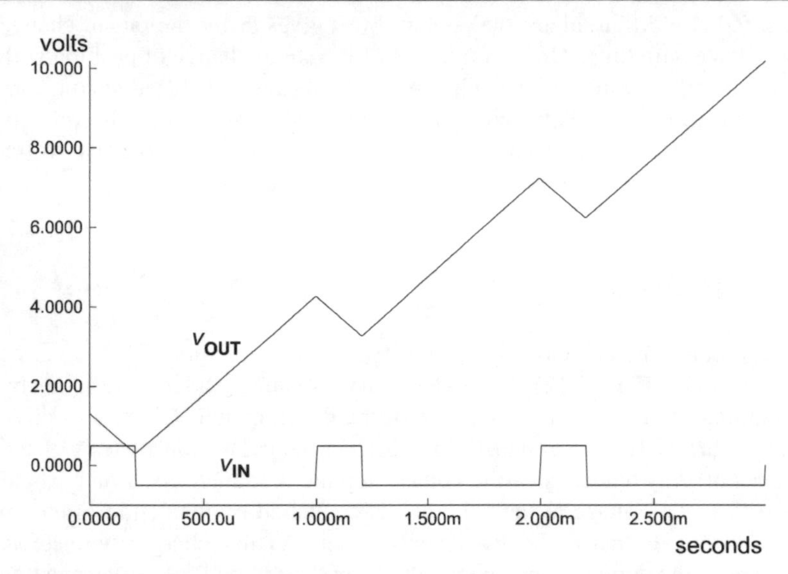

Figure 10.13 *An integrator converts a square-wave input into an upwardly-stepping ramp.*

for v_{OUT} is displaced about 1 V positive, owing to offset current in the op amp (see below). During the positive portion of the cycle v_{IN} is positive and v_{OUT} falls. Then v_{IN} becomes negative and v_{OUT} rises. But v_{IN} is negative for longer than it is positive, so the output rises steadily for longer, reaching a higher level by the end of each cycle. The next cycle shows a repeat of the short fall and long rise, with v_{OUT} climbing to an even higher level. It is clear that at any stage in the operation the inverted output voltage is a measure of the average input voltage, multiplied by the elapsed time. Mathematically the output is given by the equation:

$$v_{OUT} = \frac{-1}{RC}\int_0^t v_{IN}\,\mathrm{d}t + k$$

The negative sign indicates an inverting amplifier. The fact that R and C are in the denominator of the fraction indicate that the size of v_{OUT} is increased if either or both of these has a smaller value. The integral sign ($\int$) with the numbers and symbols around it show that v_{OUT} depends on the size of v_{IN} and the length of time elapsed. If v_{IN} varies, an average value is taken over the specified time. The '+ *k*' at the end of the expression is the initial value of v_{OUT}. This may frequently be zero.

The integrator is often used when we have a varying voltage and we are more interested in its average value over a period of time. It can be used in sensor-driven circuits. The main problem is that, if left running indefinitely, the output swings to the negative or positive rail after a very short time. Even if the average value of the input is zero, all op amps have an input offset current which constantly charges or discharges the capacitor, eventually charging it completely in one direction or the other. Fig. 10.14 shows how to counter this defect. The switch is an essential feature of all practical integrator circuits. It is closed briefly just before a run begins, to discharge the capacitor. Instead of a mechanical switch we can use an FET as a switch and so put the action of the integrator under electronic control, perhaps by sending a logical signal to the gate of the FET.

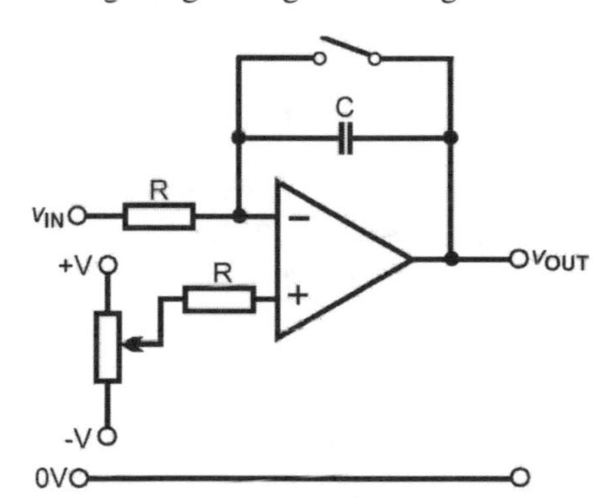

Figure 10.14 *A practical integrator has a resetting switch to start the integration from zero. It has a resistor network for counteracting the input offset current.*

If the period of integration is only a fraction of a second, the effect of input offset current during the run may be so small that it can be ignored. If the integration period is longer, we need to counteract the input offset current directly. Fig. 10.14 shows the (+) input connected to a variable resistor instead of to the 0 V line. This allows the voltage at that input to be adjusted so as to provide equal input currents at both (+) and (–) inputs and so compensate for the offset input current. In practice we would not use a variable resistor on its own as this would not allow the currents to be balanced sufficiently closely. A low-value variable resistor, preferably of the multi-turn type, with equal fixed large-value resistors connecting its ends to the power rails provides the required precision adjustment.

The circuits described above illustrate the standard ways in which op amps are commonly used. They are also used to build precision rectifiers, instrumentation amplifiers, logarithmic amplifiers, active filters, current-to-voltage converters and much other circuitry. Some of these applications are described in Chapter 11.

Summing up

An op amp has:

- two input terminals, one output terminal, two power line terminals, and sometimes offset nulling terminals,
- high input resistance,
- low output resistance,
- very high open loop gain.

The important characteristics of an op amp include:

- Input bias current, the minimum current required to bias each input.
- Input offset current, the difference between currents flowing to each input from identical sources.
- Input offset voltage, the difference between the input voltages required to produce 0 V output. It can be compensated for by offset nulling.
- Common mode gain, the very low gain obtained when the same signal is applied to both inputs.
- Common mode rejection ratio, the differential voltage gain (open loop gain) divided by the common mode gain.
- Slew rate, the maximum rate at which output voltage can swing.
- Gain bandwidth product, the frequency at which open loop gain falls to unity (transition frequency).

The basic op amp circuits are:

- Inverting amplifier, with closed-loop gain much less than open loop gain, and precisely set by resistor values. Gain = $-R_{FB}/R_A$. Inverting input acts as a virtual earth.
- Non-inverting amplifier, with closed-loop gain much less than open-loop gain and precisely set by resistor values. Gain = $(R_{FB} + R_A)/R_A$. The voltage follower is a version of this circuit useful for resistance matching.
- Summing amplifier, which is a version of the inverting amplifier with two or more signals to the inverting input. It is used for straightforward addition of signals, including mixing, or with weighted inputs for digital to analogue conversion.
- Comparator, in which the output swings fully high or low to indicate which input is receiving the higher input voltage.

- Differential amplifier, which amplifies the difference between two input voltages. Can be used as a subtractor.
- Differentiator, which has an output proportional to the rate of change of its input.
- Integrator, which has an output proportional to the average value of its input over a given time, and to the time elapsed.

Test yourself

1. Compare the use of op amps in inverting and non-inverting circuits. What are the advantages and disadvantages of each?
2. What is the most important feature of an op amp suitable for amplifying the output from a sensor with low output resistance?
3. What characteristic should an op amp have if it is used in a sample-and-hold circuit?
4. Explain what is meant by a *virtual earth*. What use is made of this property in a summer circuit?
5. Explain how an op amp can be used in a circuit for measuring velocity and acceleration.
6. Explain why an op amp integrator circuit is limited to short periods of operation.
7. An inverting amplifier based on an op amp has $R_A = 470\ \Omega$ and $R_{FB} = 22\ k\Omega$. Calculate its closed-loop gain and its input resistance.
8. A non-inverting amplifier based on an op amp has $R_A = 820\ \Omega$ and $R_{FB} = 180\ k\Omega$. Calculate its closed-loop gain and a suitable value for R_B.

11
Special-purpose amplifiers

Most of the amplifiers described so far in this book are standard types, which we have used to illustrate the main principles of amplification. In this chapter we look at some other types of amplifier which are of interest but have more specialised applications.

Instrumentation amplifiers

An instrumentation amplifier is one that is used in electronic measuring instruments or in electronic systems in which the highest precision is required. A popular instrumentation amplifier is shown in Fig. 11.1. It can be built from three op amp integrated circuits but more often the three amplifiers are manufactured already connected on the same chip. As a 'black box' the instrumentation amplifier, or 'in amp', has the same input and output terminals as a single op amp, but its performance is superior.

The signal is applied across the two input amplifiers on the left. These both have very high input resistance (100 MΩ or more) at their (+) terminals, though one of these is now the (–) input of the in amp as a whole. In Fig. 11.2, with the (–) input at 0 V and v_{IN} applied to the (+) input, the signal appearing across the outputs of these amplifiers is v_{diff}. The gain at this stage is set by selecting the value of the external resistor R_G. If R_G is 20 kΩ and the on-chip resistors are 10 kΩ, the differential gain of the first stage is:

$$A_{diff} = 1 + \frac{2R}{R_G}$$

It can be seen in the figure that the amplitude of the input signal is 100 mV, and the amplitude of v_{diff} is 200 mV. The differential gain of the first stage is 2. Alternatively, if we apply the *same* signal to both inputs, this signal appears without amplification at the outputs of both first-stage amplifiers,

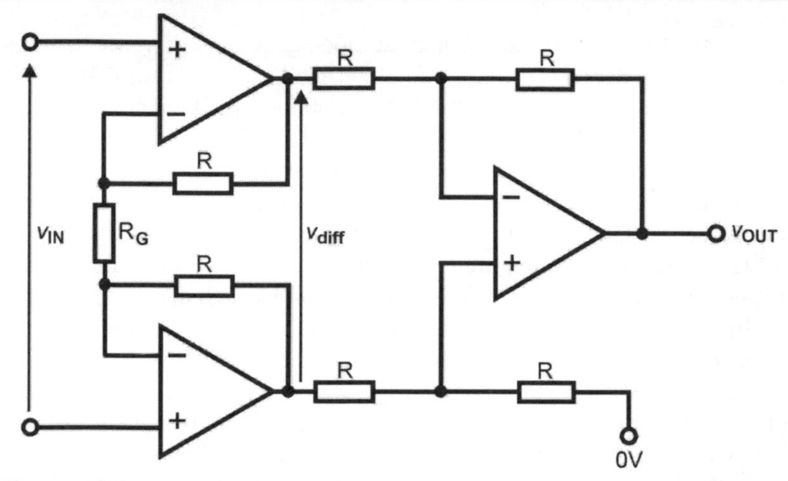

Figure 11.1 *A standard type of instrumentation amplifier comprises three op amps on the same chip, together with six resistors. The seventh resistor, R_G, is external and is used for programming the differential gain of the amplifier.*

and A_{diff} is zero. The common-mode gain of the first stage is zero.

The second stage, the amplifier of the right of Fig. 11.1, has a differential gain of 1, so the output, v_{OUT}, has the same amplitude as v_{diff} as can be seen in Fig. 11.2, but is inverted.

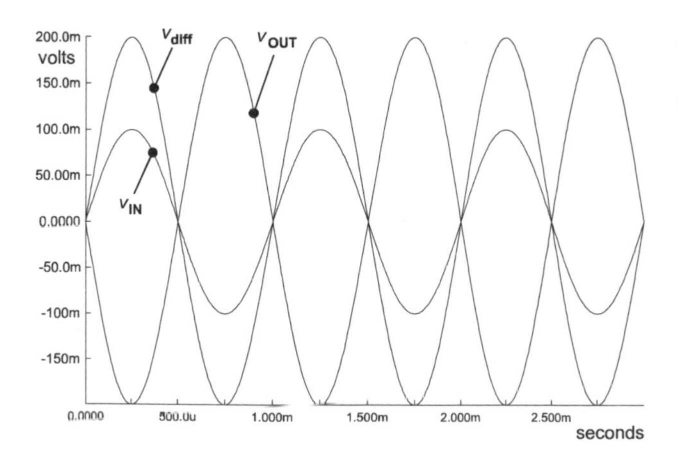

Figure 11.2 *There is differential gain in the first stage of the amplifier (v_{diff}) but none in the second stage (v_{OUT}).*

Because the first stage has zero common-mode gain, the difference between them, $v_{diff,}$, is independent of any common-mode signal from outside the amplifier. As a whole, the amplifier has no common-mode gain.

An amplifier of this kind is very useful for measuring signals from sensors such as strain gauges, pressure gauges, and thermocouples, which have a low output resistance but may be subject to common-mode interference. In medicine, this type of amplifier is used to measure voltage differences between probes attached to regions of the body in which muscular or nerve activity is to be measured. The small electrical signals are often swamped by much larger signals due to electromagnetic interference from nearby mains cables, lighting fixtures or mains-powered equipment. But the ability of this amplifier to reject all common-mode signals means that electromagnetic interference is ignored and the small signals originating in the body may easily be measured.

A related form of instrumentation amplifier is the *isolation amplifier* (Fig. 11.3). This has the same function as the in amp but it is constructed so that the input and output sides are electrically isolated from each other. They have many applications in industrial situations, such as monitoring or controlling motors, where it is often essential to separate the measuring or control circuits from very noisy circuits with high voltages and currents. The input and output sides of the amplifier are able to withstand voltage differences of a thousand volts or more and there is a high common-mode rejection ratio. There are various ways of effecting this isolation. In some types there is an opto-isolator between the two sides. In others the signal is converted into digital form and is then transmitted to the output side by capacitative coupling. The amplifier may also have a clock input for driving its processing circuits.

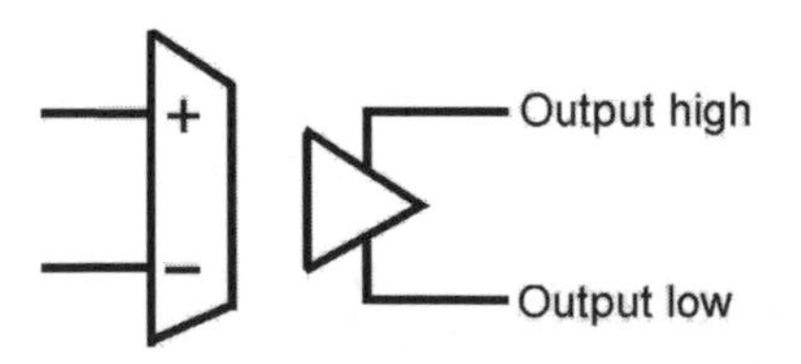

Figure 11.3 *The input and output sections of an isolation amplifier are isolated from each other, as suggested by the schematic symbol. The low output is the inverse of the high output.*

Another type of amplifier used for instrumentation where low drift is essential is the *chopper amplifier*, which is described later in this chapter.

While on the subject of instrumentation, there are a number of other applications of operational amplifiers in this field. One of these is the positive/negative amplifier (Fig. 11.4) which is useful in meter circuits when it is required to invert or not invert a signal without altering its gain. For example, if we need to measure and use only the absolute value of DC input, whether it is positive or negative, we can use this amplifier to invert negative signals so that all voltages sent to the measuring circuit are positive. When the switch is closed, the (+) input is grounded and the circuit behaves as an inverting amplifier with a gain of –1. When the switch is open, the resistances between the circuit input and both op amp inputs is 2*R*. No current can flow to the (+) input so this is at v_{IN}. The op amp acts to bring both inputs to the same voltage, with no current flowing through any of the resistors. This means that the output must also be at v_{IN}. In other words, the amplifier has a gain of +1.

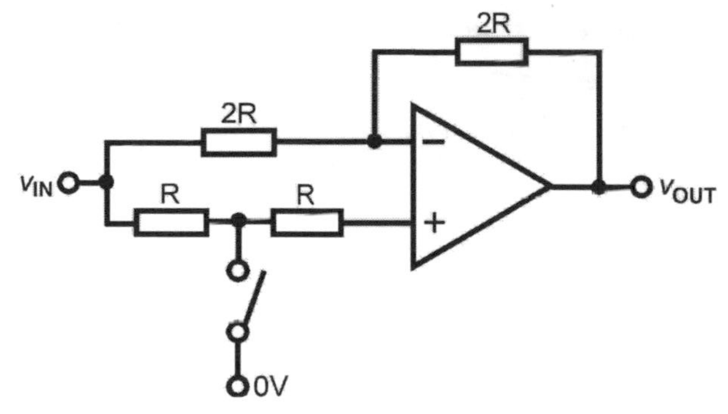

Figure 11.4 *This op amp circuit has a switchable gain of +1 or –1.*

A mechanical switch is shown in the figure but this can easily be replaced by an electronic switch such as an FET or a CMOS switch.

Transconductance amplifiers

These are a type of op amp in which the output *current* (not voltage) is proportional to the *voltage* difference between their (+) and (–) inputs. Most transconductance amplifiers also have a *bias* input, which may be used to switch the output current on or off, or to vary the transconductance of the amplifier. Transconductance amplifiers are often used as *voltage-controlled amplifiers* by applying a control voltage to the bias input. An application of this is the shaping of the envelope of electronically generated sound (Fig. 11.5). It is relatively easy to generate a tone with the correct combination of harmonics to resemble the sound of, say, a note played on a piano. But a pulse of this tone does not sound like a piano unless it has the correct envelope. The sound made when a key of a piano is struck increases very sharply in amplitude (rapid attack) as the hammer hits the string. Then after a short period of more-or-less constant volume (sustained volume), the sound dies slowly away (slow decay) as the string gradually loses energy.

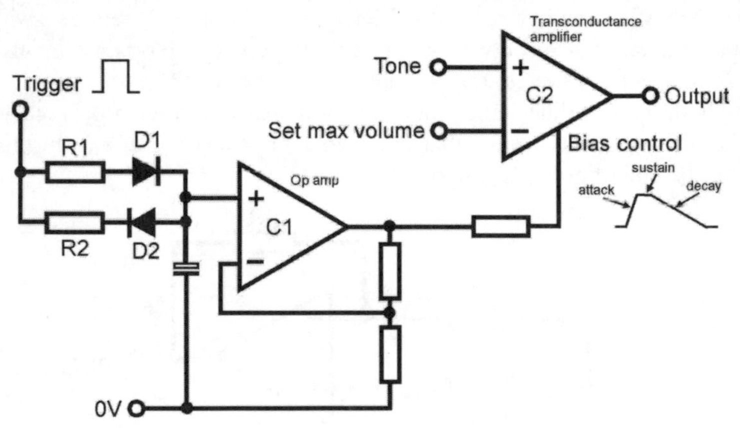

Figure 11.5 *A transconductance amplifier finds an application as an envelope-shaper in an electronic music circuit.*

In the circuit, the signal for the tone is sent to the (+) input of the transconductance amplifier. The bias control receives a signal from an op amp to produce an envelope of the required shape. The circuit to generate this signal is on the left of Fig. 11.5. The op amp is wired as a non-inverting amplifier. A trigger input, normally 0 V, goes high and charges the capacitor through R1 and D1. In the case of the piano, values are chosen to give rapid charging. As the capacitor charges, the output of the op amp quickly rises and biases the transconductance amplifier so as to produce a sharp increase in gain. The trigger holds the capacitor at maximum charge for a short sustain period. Then the trigger goes low and the capacitor discharges through R2 and D2. Values are chosen to give a slow decay. The effect is to make a sound that more closely resembles that made by a real piano.

Logarithmic amplifiers

The output of a logarithmic amplifier is proportional to the natural logarithm of the input. To build an amplifier that will produce logarithmic values, we use a component, which has logarithmic behaviour. The semiconductor diode is one such component and its logarithmic behaviour depends upon the flow of current through the pn junction. If the voltage across a forward-biased diode is v, the current i through it is given approximately by:

$$i = I_S e^{v/k}$$

I_S is the reverse saturation current of the diode, which is about 1 μA, but depends on temperature. k depends upon various physical properties of the semiconductor and upon temperature. It is about 26 mV at room temperature. So we can say that:

$$i = e^{38v} \text{ (in microamps)}$$

Or putting it in the other form:

$$v = (\ln i)/38$$

The voltage across the diode is proportional to the natural logarithm of the current. In the logarithmic amplifier (Fig. 11.6), current i flows from the input V_{IN} through R1 to the virtual earth at the inverting input of the op amp and:

$$i = V_{IN}/R_1$$

The same current flows on through the diode, which is forward biased.

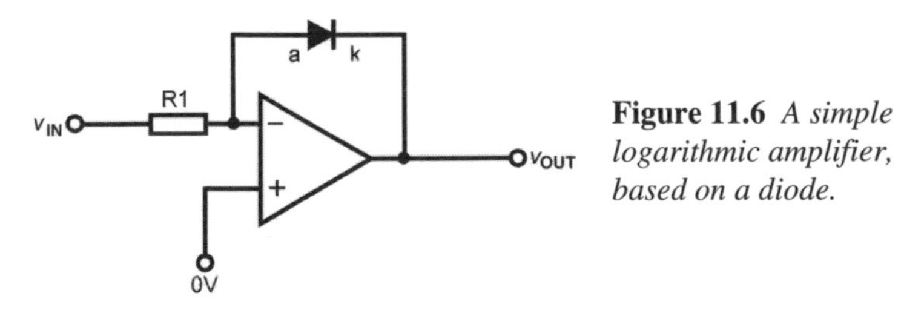

Figure 11.6 *A simple logarithmic amplifier, based on a diode.*

Therefore the voltage across the diode is:

$$v = (\ln i)/38 = (\ln V_{IN}/R_1)/38$$

Since R1 is a constant we can amalgamate this into the constant 38 and write the result more simply:

$$v \propto \ln V_{IN}$$

The voltage at the anode of the diode is greater than that at the cathode. Therefore the op amp stabilises with the inverting input v volts higher than the output. Since the inverting and non-inverting inputs are both at 0 V for stability, the output must be at $-v$. This is the output of the amplifier circuit so:

$$V_{OUT} \propto \ln V_{IN}$$

The amplifier in Fig. 11.7 uses the base-emitter pn junction of a transistor. The mechanism is the same and output is proportional to the natural logarithm of the input. This design has a few refinements. The resistor R_E is placed *within* the feedback loop of the transistor to reduce its voltage gain,

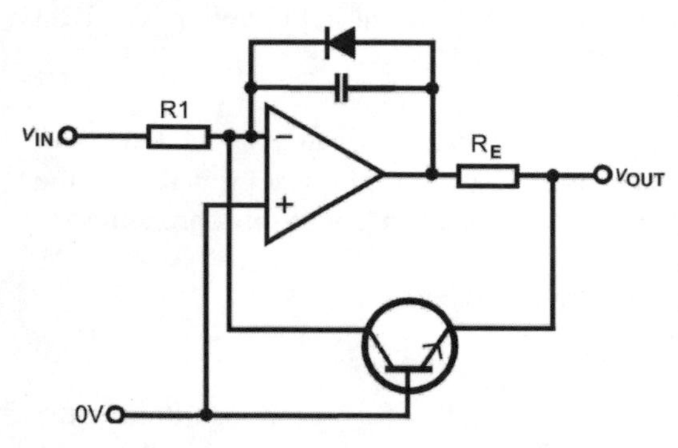

Figure 11.7 *In this logarithmic amplifier the forward-biased base-emitter junction of a transistor is used instead of a diode.*

Natural logarithms

A logarithm is an index (or 'power'). For example, we know that $2^4 = 16$. So the logarithm of 16 is 4, provided that we are talking in terms of powers of 2, the base number. We indicate the base number by writing it as a subscript to the symbol 'log', so:

$$\log_2 16 = 4$$

Similarly, $\log_2 128 = 7$, or in the other form which has the same meaning, $2^7 = 128$.

More often we use 10 as the base number, and $\log_{10} 100 = 2$, $\log_{10} 1000 = 3$. If we are working with the base number 10, we usually just write 'log', instead of '$\log_{10}$'.

In some connections, it makes equations and formulae simpler if we use the exponential constant e as the base number. This has the numeric value 2.7183 (to 4 decimal places). When we refer to logarithms to this base we use the symbol '$\log_e$' or more often the shorter symbol 'ln'. Logarithms to the base e are referred to as *natural logarithms*.

which might otherwise lead to oscillations. The diode, which is oppositely oriented in this figure, prevents the op amp from being over-driven by negative input voltages. The capacitor is there to improve the stability of the circuit.

Logarithmic amplifiers are used in measuring circuits in which the quantity being measured varies logarithmically. An example is the measurement of sound levels. These can vary over a very wide range and we usually express them in decibels. Decibel readings ranging from 0 dB to 100 dB could represent a range of power from 1 to 10^{10} watts. Or it might sometimes be milliwatts, for a decibel scale measures ratios, not absolute units. In fact we can choose any value to be equivalent to the 0 dB level.

Fig. 11.8 shows the output of the amplifier of Fig. 11.6 while the input signal is varied logarithmically over the range 10 mV to 10 V (a thousand-fold range). The output signal falls steadily and it needs only the eye to tell

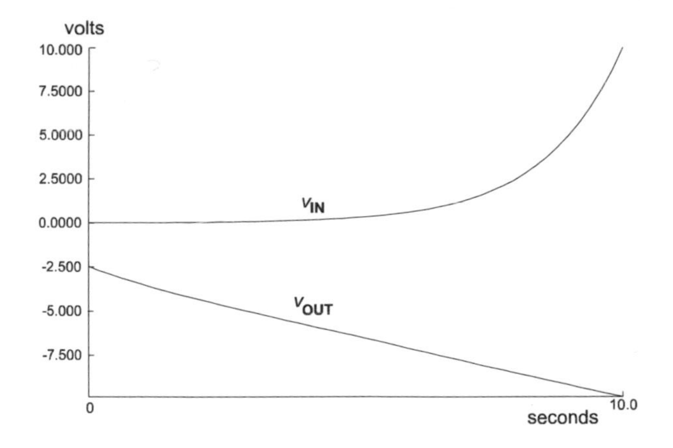

Figure 11.8 *The input to a logarithmic amplifier is made to increase logarithmically with time. The output falls steadily. Its graph is clearly a straight line.*

us that there is a straight-line relationship between the output and the logarithm of the input. If such values are plotted on a linear scale it is very difficult to judge whether the relationship is actually logarithmic or whether it has some other non-linear dependence. Converting a logarithmic function into a linear one enables us to tell at a glance if the plot is linear (and therefore the relationship is logarithmic) and to measure its slope.

Comparators

A comparator is used to compare two input voltages and indicate by its output which is the higher. An ordinary op amp can be used as a comparator (see Fig. 10.8) but better performance is obtained from integrated circuits specially designed as comparators. Comparators have a higher slew rate than op amps so that they have a quicker response to unequal inputs, which is important if the comparator is to operate alongside fast logic circuits.

Normally a comparator is operated in an open-loop circuit, like that used for the op amp comparator, making use of the very high gain of the comparator to swing the output one way or the other as soon at the input voltages differ. The main problem with using comparators is that, there is nearly always a small amount of stray positive feedback in the circuit layout. For example, there may be capacitance between adjacent tracks on the printed circuit board. With positive feedback, the comparator may oscillate as its output swings from positive to negative, or in the opposite direction. If the output of the comparator is being fed to a logical counting circuit, the oscillating output could be taken as multiple counts. The oscillations also generate radio-frequency interference, which may be troublesome. Oscillation may be avoided by designing the layout of the circuit with care to avoid stray positive feedback. Another approach is to deliberately introduce positive feedback (Fig. 11.9) to swamp the effects of any stray feedback that is present. This introduces *hysteresis* into the action of the circuit. Output changes state from negative to positive when the voltage at the (+) input exceeds that at the (–) input by a certain small amount. Once this has happened, the voltage at the (+) input must fall a certain amount below the voltage at the (–) input before the output can change back to negative. In other words, the threshold voltages for the two

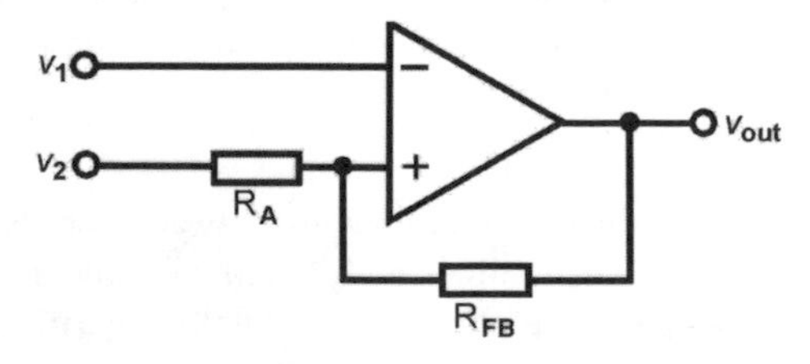

Figure 11.9 *One way to avoid oscillation in a comparator is to provide positive feedback.*

directions of change of state are not the same. Once the state of the circuit has been triggered in one direction, small reversals of direction have no effect. The circuit flips positively from one state to the other, without oscillating on the way.

Another way of avoiding oscillation is to use a *clocked* comparator. This samples the input voltages at regular intervals, as determined by the built-in clock, and then changes the output accordingly. Once the instant of sampling is past, any positive feedback that may be present can have no effect, for the output can not change until the next sampling instant.

Comparators are made with various types of output, some of them specially suited for interfacing to particular kinds of logic circuit. The *open-collector* output requires a pull-up resistor, to take the place of the collector resistor missing from the output transistor stage (Fig. 11.10a). This has the advantage that the pull-up resistor can be connected to the positive rail of circuits operating at a higher voltage than that of the comparator circuit. This makes interfacing to a wide range of circuits very easy. The main point to remember is that, if the following circuit has low input resistance, it may take an unacceptably long time for the output voltage to be pulled up to a high level. To avoid this, the output from the comparator should be fed to a buffer (for example, an op amp wired as a voltage follower) which has high input resistance.

The *totem-pole output* (Fig. 11.10b) needs no pull-up resistor and has the advantage of active pull-up which gives it a rapid response time. Comparators with totem-pole outputs are generally made for interfacing to logic

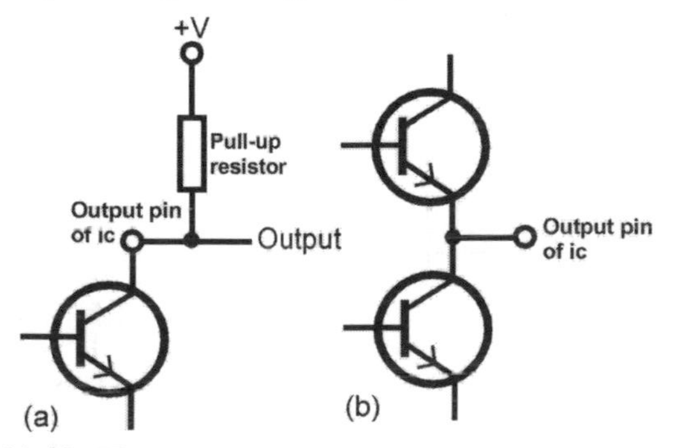

Figure 11.10 *The output of an integrated circuit comparator is usually either (a) open-collector, or (b) totem-pole.*

circuits operating on the conventional 5 V supply. They are not suitable for use with other supply voltages.

Chopper amplifiers

Even the best of the instrumentation amplifiers described earlier in this chapter are subject to a certain amount of input offset voltage error and drift. To make matters worse, these errors are dependent on temperature. This means that temperature changes have to be compensated for continually. The chopper amplifier is intended to minimise these errors by monitoring them automatically and then correcting for them. The term 'chopper amplifier' has been applied to a number of different types of instrumentation amplifier, all of which are intended to be drift-free and offset-free. Here we describe one of the more recent designs, also known as an *auto-zeroing* amplifier, or *chopper-stabilised* amplifier. It is available as a complete unit on a single chip, except that the capacitors C1 and C2 are usually external.

The amplifier consists of two op amps connected as shown in Fig. 11.11, together with an on-chip clock running at a few hundred hertz. Although there are two amplifiers, the connections to external circuitry consist of

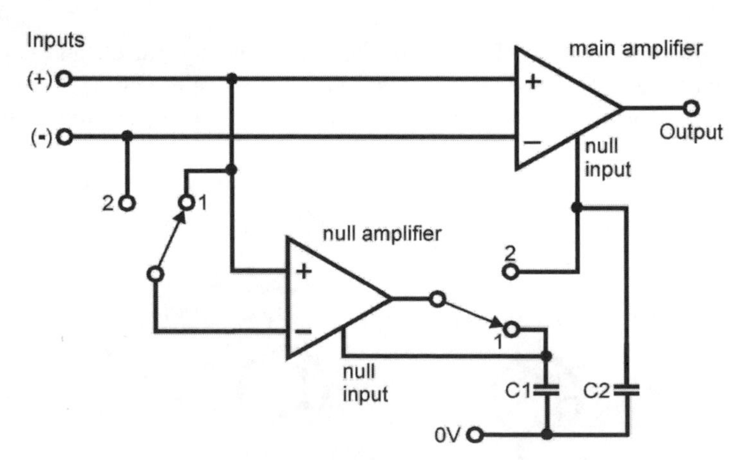

Figure 11.11 *A chopper-stabilised amplifier includes a null amplifier, which is used to correct its own errors and those of the main amplifier.*

only a (+) input, a (–) input, a single output and a connection to the 0 V line. So the complete chopper amplifier can be wired into standard amplifier circuits including inverting and non-inverting amplifiers, comparators, summers, logarithmic amplifiers and others. With on-chip capacitors for the clock, the chopper amplifier can be treated just like an ordinary op amp.

The two amplifiers have different functions. The signal is amplified by the *main amplifier*, while the *null amplifier* is used to detect and correct the errors of both amplifiers. The clock operates two changeover switches, which are really based on CMOS or other transistors but are shown as mechanical switches in the figure to make their operation clearer. In the first stage the switches are both in position 1, as shown in the figure. The input terminals of the null amplifier are shorted together and its output is connected to C1. If the null amplifier is already corrected, and with both its inputs at the voltage of the (+) input, its output should be zero. Otherwise, if uncorrected, it is at a slight positive or negative voltage. The other switch connects the output of the null amplifier to C1. This stores the output voltage, which is also fed to the null input. There it operates to correct the null amplifier for input offset voltage and drift, so that its output becomes zero. This completes the first phase of operation.

In the second phase of the clock the switches are in position 2. If the amplifier is operating in one of the standard op amp circuits, its two inputs (+) and (–) are at equal voltages, and both the main amplifier and the null amplifier have equal inputs, since they are connected in parallel. However, if the main amplifier is uncorrected for offset and drift, its input voltages are unequal and so are those of the null amplifier. The output of the null amplifier has a small positive or negative voltage, which is fed through the switch and stored on C2. It is also fed to the null input of the main amplifier. This applies a correction to the main amplifier, until its inputs are of equal value.

With the errors in both amplifiers now corrected, the amplifier as a whole (that is, the circuit as drawn in Fig. 11.1) behaves as a precision op amp. Its inputs have been brought to more-or-less equal voltages (but suitably corrected) and it output is of the correct value expected from a high-precision amplifier. The performance of the amplifier is superior to that of any high-precision bipolar op amp. The main disadvantage is that the output carries high-frequency spikes from the clock. These can be ignored or filtered out for low-frequency applications but limit the usefulness of the amplifier at high frequencies.

Parametric amplifiers

We have left one of the more unusual types of amplifier until last. As their name suggests, parametric amplifiers depend on altering one of the *parameters* of the amplifier while it is running. The parameter that is altered varies according to the design of the amplifier, but in the example given here the altered parameter is the capacitance of one of the capacitors. Before considering how this is done, let us return to the analogy of the child on a swing. In previous discussions of this kind we have referred to supplying energy to the swing by pushing it at suitable times during its cycle. But a child can 'work' the swing without outside assistance simply by standing up on the seat and sitting down again at the right times, twice during each cycle. Standing up and sitting down alters the effective length of the swing for it raises and lowers the centre of gravity. The length of the swing is one of its parameters and the child decreases and increases it at suitable intervals This is an example of the conservation of angular momentum. The child does work by standing up, and this appears as an increased

Parameters

The word 'parameter' is frequently misused. It originated as a mathematical term but is also applicable to electronics. In electronics, a parameter is a quantity that is normally held constant in a given circuit running at a particular time, but which may be different when the circuit is run at another time. For example, the power supply of an amplifier may be constant at 9V on one occasion, but may be changed to a constant 12V on another occasion. The voltage of the power supply is a parameter of the amplifier. Contrast this to the signal voltage, which varies continuously while the amplifier is running. Signal voltage is not a parameter. Similarly, the voltage gain of a BJT amplifier is not a parameter because it may vary during the running of the circuit according to the collector current.

In the example of a variable capacitor, its parameters are the area of the plates, their distance apart, their area of overlap, and the dielectric constant of the material between the plates. The charge on the plates and the voltage between them are not parameters.

A parametric amplifier amplifies waveforms of a particular frequency, usually in the microwave RF region. Fig. 11.12 illustrates one type of parametric amplifier. The signal is fed to one winding of a transformer. Another signal, the pumping signal, is fed to a second winding. The pumping signal is a square wave of exactly double the frequency of the input signal and is synchronised with it. It is applied across a special kind of diode known as a varactor diode. When reverse-biased, the diode has capacitance between its cathode and anode, in the same way that we have seen capacitance to exist between the base and emitter of a BJT. Here we make positive use of this. The charge-free depletion region acts as the dielectric. The thickness of the depletion region depends on the reverse bias across the diode, and the capacitance is inversely proportional to the thickness of the depletion region. Diodes such as this are often used to adjust the tuning of LC networks in radio frequency circuits. In a typical varactor diode, the capacitance may be varied from 10 pF to 160 pF varying the reverse bias from 10 V to 1 V. By subjecting the diode to a square-wave pumping signal we can alter its capacitance while the circuit is running.

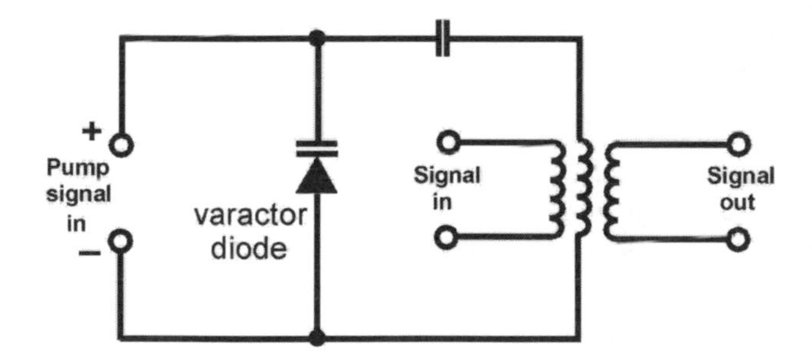

Figure 11.12 *This circuit operates by decreasing the capacitance of the varactor and so increasing the signal voltage amplitude.*

The charge q on a capacitor of capacitance C, is related to the voltage v between its plates by the equation:

$$v = q/C$$

This shows that we can increase the voltage across the capacitor by decreasing the capacitance, the charge on the capacitor remaining unaltered. In other words, we can amplify the voltage by altering a parameter of the capacitor. The energy to do this comes from the pumping circuit. This is equivalent to the child standing up on the swing, using energy and so increasing the energy in the swing.

The pumping circuit decreases capacitance (amplifies voltage) as it swings negative. Later it swings positive again, which would simply increase capacitance and restore the voltage to its former level were it not for the timing of the pumping circuit relative to the input signal. We arrange that the pumping signal swings negative at exactly the same instant as the input signal reaches its highest positive amplitude. This boosts the voltage to a higher level. Because the pumping signal has twice the frequency of the input signal, the pumping signal swings positive just as the input signal passes through zero. If it is zero, the change of capacitance has no effect. The input signal then enters its negative phase and is pumped to a more negative voltage as the pump signal swings negative again.

In this way we increase thc voltage across the capacitor at both peaks of its cycle and the input signal (now amplified) is picked up by a third winding of the transformer to become the output of the amplifier. One of the difficulties of operating this circuit is that of keeping the pumping signal synchronised with the input signal.

As an alternative to the varactor, some parametric circuits use a capacitor with four electrodes. The dielectric is a special ferroelectric material. The dielectric constant of this depends upon the voltage between the two control electrodes. In Fig. 11.13, the signal is applied to the control electrodes so varying the capacitance in the output circuit. A generator in the output

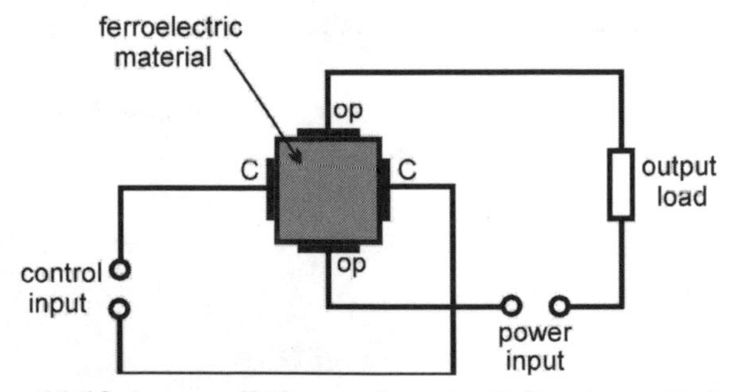

Figure 11.13 *A controlled capacitor, its dielectric consisting of a block of ferroelectric material, is the heart of another parametric amplifier. C = control electrodes, op = electrodes of the output circuit.*

circuit supplies power to produce a sizeable current through the load. The current in the circuit (and hence the voltage across the load) varies with the capacitance, which is under the control of the input signal.

One of the chief advantages of parametric amplifiers is that no resistors or semiconductors are involved in amplification. This means that there is no source of noise in the amplification process. For this reason, parametric amplifiers are used in radio astronomy and in satellite communications where signals are very weak and the lowest possible noise levels are imperative.

Summing up

The special amplifiers described in this chapter are:

- *Instrumentation amplifiers* High-precision amplifiers with gain programmable by an external resistor. Reject all common-mode signals so are ideal for measurement of small voltage differences against a background of interference and noise.
- *Isolation amplifiers* Instrumentation amplifiers in which the input and output sections are electrically isolated from each other.
- *Positive/negative amplifiers* Amplifiers in which the gain can be switched between +1 and –1 under electronic control.
- *Transconductance amplifiers* In these the voltage difference between the inputs produces a proportionate output *current.* A voltage applied to the control terminal determines the transconductance of the amplifier.
- *Logarithmic amplifiers* Their output voltage is proportional to the logarithm of the input voltage.
- *Comparators* These are similar to ordinary op amps but have very high slew rate. Their outputs are open-collector outputs for interfacing to a wide range of circuits, or totem-pole outputs for interfacing to TTL and other 5 V logic families.
- *Parametric amplifiers* Their action is based on varying a parameter of a capacitor or other component so as to 'pump up' the input signal to a higher amplitude. They are used in applications where noise levels must be as low as possible.

Test yourself

Select one or more of the types of amplifier described in this chapter. Write a concise description of its circuit (including a circuit diagram, if possible), how it works, and in what application(s) it is used.

Appendix

Answers to numerical problems

Chapter 1

Test yourself
5 (a) 750 Hz.
5 (b) 1.5 kHz.

Chapter 2

Keeping up?
3. 0.12 A.
5. 11.75.
6. 3.65 V.

Test yourself
7. 680 kΩ and 330 kΩ, 222 kΩ.
8. 3.58nF.

Chapter 3

Keeping up?
1. 1, 0.997.
2. 1.25 Ω.

Test yourself
3. 2.996 V.
4. 5 Ω.

Chapter 5

Test yourself
6. 58.75 kΩ in parallel with the biasing resistors.

Chapter 6

Test yourself
4. 16 000.

Chapter 7

Keeping up?
6. –100 dB.

Test yourself
2. 1.5 kHz and 3.5 kHz.

Chapter 10

Test yourself
7. –46.8, 470 Ω.
8. 221, 816 Ω.

Index